The Blackbird's Tale

or

How an aircraft Suckered me into things I never believed possible with the help of an Aircraft Carrier!

By

Jeannette Remak
(aka Det 1)
(©Phoenix Aviation Research 2017)

Acknowledgements

I would like to thank all my friends and co-workers for all of their support during this phase of my life in which this book takes place. Their names are withheld to protect the innocent and not so innocent.

Please note that there are names, which for various legal reasons, we chose to use an alias to protect them and us. However, the story is as true as it can be.

Forward

I have been through a lot in my life to say the very least. However, what you will read on the following pages is the start of a career. Messy though it was, this chapter in my life led me to the place that I am now, and that is a military aviation historian. I learned a lot from this little episode. What makes this episode important is that it started me on the road to where I didn't know I wanted to be! I found it out the hard way because it grabbed me by the scruff of the neck and said, "this is your place!" Lots of us start out after college having a pretty good idea of where we want to be, well to be sure, this is the LAST place I thought I would ever end up.

I started my life, per say, as an artist. I went to college to become a fine artist. I also went to college to become a specialist in commercial photography, which led me to the digital world of photography when things changed. The Photo/digital industry was very good to me. It gave me the means to do things I would have never thought possible.

In between all of that, I was bitten by the aviation bug. Man! Did it bite hard!!!! I never knew I could love something so much! I love my art, sure, but writing became my world, although I still paint with a mean brush, aviation art, by the way. Yes, I still like to paint still life, but I like it better when it smells like hot metal and reeks of oil and JP-7.

I learned that airplanes have a story to tell. Yes, each one has a unique story. Their history is there in the metal, in the wings, in the dusty cockpits of aircraft in museums. It's there, and it's strong. It burns your fingers the moment you touch them. The aircraft want to tell you what they have been doing. Sure, it sounds nuts, I know, but it's true and anyone that loves airplanes knows this to be true. While I did have to blank out some names, the ship the EVIL I, is a lady I love dearly. I am calling her the "EVIL I" to make sure the story can be free from any libel, and at least now, I can say she is a good museum. When I was there, problems were many and they were tangled. However, with a lot of love, and burned fingers, I managed to get the job that was asked of me done, and a lot more with it. I loved every second and it brings me right here to today.

I never thought I would ever release this story, but it's been 20 years and things along with times have changed. This is a good story because it shows you never know what tomorrow will bring….. you never know! And you never know what will capture your heart and make you want to go on, no matter what. That's exactly what happened to me.

Contents

List of Terms -- pg. 8

Chapter I – Hello ….my name is.. pg.10

Chapter II – How I became a spy for the USAF Pg.12

Chapter III – The spy who came in from the cold…er..crosstown bus. Pg.27

Chapter IV – Blackbirds and Procurement Pg.37

Chapter V – *Evil I* Traps the Spy Pg.56

Chapter VI – Blackbirds, grunge work and dancin' on the edge. Pg.70

Chapter VII – Buddies, Airplanes and Presidential nightmares Pg. 91

Chapter VIII – What helicopter? Pg.119

Chapter IX – Road trip to find a missing bird Pg.123

Chapter X – Threats and a change of command Pg.128

Chapter XI – The end of the line and the on set of Ravens. Pg.138

Chapter XII – #122 Pg.141

Epilogue – Today Pg.142

List of Terms

JV-- partner who worked with me

GH --Member of the USAF Museum office

TB--Member of the USAF museum office

SC--member of the USAF Museum office

PC-- Program Office USAF Museum

MR--NASA/USAF friend

GP--NASA office /friend

GJ--someone we knew who liked Blackbirds

LD--friend from NASA

Navy Z--Friend from the NAVY

Ace-- worker on the Evil I in charge of Ops

Buddy--worker at Evil I

Lou--Worker at Evil I

Museum Director--fairly easy to figure out

JR--Friend from one of the Aircraft companies we worked with

My Buddy #1--- my first friend and associate at Evil I

My Buddy #2--my second friend and still associate from Evil I

My Boss--the supervisor of the flight deck maintenance crew

D---Member of the Evil I Op department

MT--worker on Evil I

MW--Naval Historian

PL-- Associate from the General Administration Office

AV--GSA office NY associate

HD—former crew member

WB--Member of the Evil I office

General S--first General who took over Evil I Administration.

This is the VERY first time that I met #122. That is her right by the maintenance handler. (Author Collection)

Chapter I

Hello ... My name is....

Allow me to introduce myself, my name is "SUCKER". A 100%, full fledged SUCKER. What type of sucker you say? Well, probably the best kind. While I may have been just that, a sucker, I fell into it for the best of reasons. I tried to save something precious to Aviation and American history. I fell in love with the most exotic, demanding airplane you could ever imagine. Not only that, I fell in love with the ship that held her captive. Strange tale? Yep, you betcha! How could a five foot three inch tall, redhead born and bred NYC woman, with a mild interest for airplanes and aviation history end up falling in love with the A-12 Blackbird and an aircraft carrier that changed her life to where she's written many books and created many paintings on military aviation history? Yes I agree, I haven't figured it out myself just yet and it's now been a while!

However, the people that know the ins and outs of this saga have compelled me to put it down on paper. I guess because they know no one would believe them if they actually told this story. I will admit that the six years that I was involved with the Aircraft Carrier Museum *Evil I* and the National Museum of the USAF's™ aircraft, the A-12 Blackbird, were six of the most intense years I've ever spent doing anything!

Regardless of the pain and suffering endured for both these beloved pieces of U.S. military/aviation history, I wouldn't change a thing, not ONE THING! I certainly made a difference for the better. Perhaps that's all you can ask for in any situation.

Because of the strange nature of this tale, anyone familiar with the famous "Pirate's of the Caribbean "Curse of the Black Pearl "will understand the choice of music for this saga…. I will now begin the "Blackbird's Tale".

Chapter II
How I became a Spy for the USAF

Before we get started, I really wanted to call the ship her exact name, however, because of legal issues, I don't wish to be involved in, I decided to use one of her nicknames. The reason for the nickname comes straight from her crew members..." If she loved you, she would protect you.... If she hated you.... She'd kill you." Let's face it, in all the wars and action that the *Evil I* has seen, that message from the crew was basically prophetic when looking at her battle record. When I first met *Evil I*, I was like every other born and bred New Yorker. I'd never seen an aircraft carrier up close and here was one parked on 12th Avenue and 46th Street, right off the Hudson river.

I remember the story that appeared in the Daily News vividly. *Evil I* had just come into possession of a "Blackbird". While I was familiar with the SR-71 and not that familiar with Aircraft Carriers, I knew I had to see this. Besides, where else in New York City could you come in contact with real, honest to goodness warbirds? My interest for airplanes, particularly big jets and military aircraft was getting the better of me lately, and it was an itch I just had to scratch.

So, hauling my ever suffering, best friend/cousin out to 12th Avenue and 46th Street, we paid our admission and I was formally introduced to the *Evil I*. Little did I know that very soon she would take possession of me, body and soul, intensely and sometimes painfully, but most of the time joyously.

We climbed up those stairs to the flight deck and just as we stepped out of the staircase, to the left was the object of my soon to be affection. Talk about love at first sight! I couldn't get up that maintenance handler fast enough to get right by her side. The Lockheed A-12 Blackbird just seemed to be moving while standing still. Oh! those incredible nacelles, the long fluid lines, those lovely knock-kneed struts. Yes, knock-kneed. When you look at a Blackbird from sort of an up angle and you see those wheel well doors low to the sides of the tires, it gives a knock-kneed appearance. The trouble really stated when I finally did get up that maintenance handler. I need to interject something here. It's very important to the story and the understanding of what went on. I am one of those individuals that can sense things through touching them. Basically, what that means is I can pick up impressions from objects. I'd been aware of being able to do this since I was a kid. My Dad used to take me to used car lots on Saturday afternoons. We loved to go looking at all the cars. He would make me touch the different cars for sale that he was interested in. I would tell him if it was a lemon or if it was good.

I was never wrong. He made the mistake of not listening to me once and bought a really pretty Dodge. Within three weeks the car was up on the rack with transmission problems that never were corrected. I warned him, and he bought the car anyway. Well to say the least, the car was sold about two months after he bought it. I told him I had a bad feeling. He never bought another car without me being there. Well, this magic intensified when I got around airplanes. For some reason, airplanes talked to me in an imperceptible way. So much so, that I could usually get close to whatever the problem was that was bothering them. This puts us precisely at the moment when A-12- #606925, CIA Article #122 and I met. Let's put it this way, I melded much like the way Vulcans do. If you are a fan of Star Trek, you know just what I mean. I have to tell you that my first touch with #122, the A-12 was a sharp tingle. I remember it distinctly. When my hand touched the skin of her fuselage, just below the cockpit, I felt that this girl was upset. Another thing, my business partner and I developed throughout our long friendship and even into our writing, that standard that aircraft are known as SHE. Neither of us plays into the political correctness thing. Ships and airplanes are called "she". As with all females they carry and nurture life.

Why should it be any different for the machines that serve us in the most critical and dangerous times of our lives, most especially for our military? I was confused at first, for I really didn't understand the nature of a Blackbird, but that would eventually come in time. (Oh Brother! Would IT!!) I went down the maintenance handler, not because I wanted to but because there were about a hundred people behind me waiting to get up there. When I finally stepped off, my cousin said to me" You should see the look on your face. It looks like you were struck by lighting". Actually, there is an existing photo of that moment that she snapped without my knowing it. It was true. It wasn't until I started to walk around #122, that I started to realize what the problem was. All these people, pawing her, grabbing, pounding, banging on her fuselage, tugging at her inlets and pitot tube. That's what was wrong. This aircraft never had anything but the familiar careful handling of her crew chief and pilots. You must remember that this aircraft, #122, was actually the flight test aircraft for the fleet of A-12s that were hidden away in the desert of Nevada, in the infamous place known as Area 51.

#122 was the first "X File". She'd never seen so many people together, let alone sitting 17 stories up on a ship floating in New York Harbor. It sort of reminded me of a young starlet from one of those 1950's Hollywood premieres.

It was really scary when you thought about it. I started to examine the aircraft from the wheels up. I also started to see a multitude of problems, but then again, if you looked at the other aircraft on *Evil I*'s deck, you knew that top maintenance and restoration was not a big issue with them. It was a shame, too. There were some really fine aircraft up there. With a little love and attention, all the aircraft could be spectacular. Well, I made a mental note of just how #122 stood on that spring day, patted her on the belly and told her I would be back to see her again. Was that ever an understatement!

The story now changes scene to the National Museum of the USAF ™ at Wright-Patterson AFB in Dayton, Ohio. It also brings us to my life in writing, espionage, and general annoyance of the status quo. My business partner, co-author and I actually met over another aircraft, the XB-70 Valkyrie Bomber.

Possibly another time would be better to describe exactly how that came about. Suffice to say, Valkyrie actually set into motion the situation that was going to occur at the *Evil I*. The National Museum of the USAF ™ was actually the owner of the A-12 Blackbird on *Evil I*. The aircraft had actually been destined for another museum on the east coast. The director of that museum had actually set aside $250,000 in an escrow account to support the move of his A-12 Blackbird #122. This museum could have offered #122 a classic home inside the museum with the right kind of display and special care that an aircraft of her stature needed.

But something was afoot concerning #122. The USAF PC office (program coordination) was in charge of all aircraft in its inventory. That basically meant that if a surplus aircraft came up, Museums would then put their bid in. If they met the requirements PC set out, the aircraft was theirs. Here's the twist.

At the time #122 would come up for bid, everyone wanted an SR-71. The A-12, being the more rare of the birds, was put out there when there were no more SR-71s. When the bid was put out, this other museum was the first in line. That was until someone else came into the picture. Enter the very rich, equally " I wanna be in the Military and can't be", New York real estate magnate. I would LOVE to trash him into the next millennia, but I can't. If he did one, sole thing right, for all of his business deals, it was bringing the *Evil I* home to New York, thus saving her from the scrappers. In the case of the A-12 however, ole' real estate magnate was determined that *Evil I* would have a blackbird. It didn't matter that the Navy never owned one and didn't belong there, as far as history was concerned. He wanted it. It didn't matter who or what he stepped on to get it.

Which is just what he did. To put it plainly, the real estate magnate "bought" the A-12 #122 out from under the very professional Director of the other museum. It was the real estate magnate's "monetary influence" with the upper echelon of the USAF that put one of rarest of airplanes in jeopardy. This real estate magnate pulled enough strings at the Pentagon to swing the deal. Of course, donations to the USAF in the form of a project known as " H----- House" helped to seal the deal. Just for the record, a "H---- House"is based on the concept of the Ronald Mcdonald house, used for sick or terminally ill children for medical care and hospice service that allows the family to live at the facility with the child. However, the military version was open to all servicemen and their families, with not quite the same concept. While I have no real issue with the concept of a "H----" house, I do take real issue when it was used as a bartering tool to gain airplanes for Museums that weren't up to the standard for preserving and protecting rare aircraft. Face it, historically the *Evil I* is the very last place that an A-12 should be, even today some 20 years later. The A-12 never served in the Navy, never flew off a carrier deck and most of all, being 93% titanium, had no business being out on an open flight deck in all kinds of extreme weather! Well, that didn't matter to the real estate magnate. He managed to get his aircraft. So much for the fundamental acquisition policies of a Naval Aviation Museum, and believe me the Navy wasn't real happy about it either.

The" ill gotten" #122 made her way up from the southern U.S. via a barge, which was being towed by one of *Evil I* tug boats. This was a rather unique way to move an aircraft made of 93% titanium, which due to the chemical interaction of salt water and titanium, already influenced the corrosion that she later suffered and continues to suffer today. So, picture this, a bunch of wannabe Indiana Joneses are now moving an aircraft they know nothing about, haven't got a rat's ass worth of a clue how to properly prepare for shipment, put her in a situation detrimental to her survival and think they're "HOT" doing it! Total Morons. The least they could have done was leave her in *spray lat*, which is sprayed on latex covering.

This is the only way to describe the *Evil I* crew that picked up #122, and to be quite honest,
 I blame the USAF and the National Museum of the USAF™ PC office for caving in and letting this debacle happen.

The A-12 on a barge on the way up from down south. (Author Collection)

Yeah, I've seen the photos of the move and to be totally blunt, it makes me want to puke and then toss it right into the arrogant faces of the *Evil I* "Ops" department that orchestrated this horror of a move. I can only imagine the rest of the pathetic trip, and actually it hurts too much to do that, so I will pick up the journey as she enters NY harbor.

A-12#122 arrived much like the Queen Elizabeth II did, along with the other big ocean liners and berthed right across from *Evil I* at Pier 88. From the videos and photos that I've seen, the aircraft was crane lifted onto the flight deck of the *Evil I*, suffice to say that this band of wannabe's proceeded to reassemble the aircraft BACKWARDS!!!!!

Since they were provided with no information, the manuals were non existent, at least at that time, and PC didn't give a hoot what they did, they installed everything wrong which meant that they then were stripping rivets and tearing holes to take stuff off and try it another way. They didn't even BOTHER to hose her down with distilled water to remove the salt spray from her, but most of all.......they NEVER attempted to ASK anyone if in fact they knew something that would help them. Hence, my readers is the role of arrogance in the movement of rare airplanes. Well, #122 being the trooper that she is held together despite their arrogant, insensitive, and cavalier treatment of her. It was really amazing that she did and that was due to the monumental job of construction that the Lockheed Skunkworks and Kelly Johnson did.

We now find the A-12 firmly settled on the flight deck of the *Evil I*, left open to the ministrations of the wannabe's who attempted to sand the old paint off her, repaint her with the best of Sears house paint, paying no mind to that bright green corrosion forming around her rivets. Of course, standing the aircraft's full weight on struts that were pumped up beyond the requirement, sitting on rims of flat tires, which also didn't help things out too well, either. However, the brilliant idea for displaying the aircraft that way was you would be able to see her better from the flight deck, when on the ground looking up from 12th Ave. and Westside Drive. Bright boys they were on the Evil I, anything for a good side show. I bet you are ready to ask why would I become involved in this madness? Good question. I haven't got an answer, but get involved I did.

We will now venture into the realm of lunacy. This is total lunacy that gripped me and took six years of my life. It still continues today but in a much quieter, less physical way. Since that first day I saw #122, I knew she needed help. It just wouldn't leave me alone and thanks to **JV**, my partner and curator at the National Museum of the USAF, and Valkyrie, #122 would get that help she needed so badly.

Let's start at the almost beginning. I had been traveling back and forth to the National Museum of the USAF (TM) to see **JV** and XB-70Valkyrie. **JV**, then in research for the Museum and I had been talking about the A-12 and the poor condition she was in. Every time I went to *Evil I* and saw her, I would continue to tell **JV** how badly she was doing. Before we knew it, we were keeping notes on her status. Talk went back and forth between us and finally the "Program Coordination" office and its then boss GH. Apparently the relations between GH's' PC office and the *Evil I* were anything but cordial. *Evil I* had been giving them the "Okie dokie" treatment for sometime concerning the A-12 . That was because they had not only contempt for the PC office, but they felt that they were above anyone telling them what they had to do with "their" aircraft. Yep, you guessed it, they never read the contract that said the aircraft was on **LOAN** to the *Evil I*, they didn't own it. As a matter of fact, they didn't care at all that every aircraft on that deck had a contract from the Marines, Navy, Army or Air Force. Well, let's just say almost every aircraft on the deck. Some exhibits even then were more than a little shady. Things did have a tendency to come and go at *Evil I*.

The trips to the *Evil I* on my part were becoming a once a week thing, with a follow up phone call to **JV**. Trips to Ohio were also becoming more and more frequent. That's when it was easy because US Air had direct flights there due to Dayton International airport being its hub. Trips were usually a one day turnaround. I would leave for the airport at 6am, reach Dayton by 9:30 , get out to **JV** and the Museum , spend the day and be back at the airport for the 7pm flight to NYC and home by 9pm. That's if all went well and there were no Nor'easters to hold me over in Dayton, which did happen on occasion.

Well, somehow and I don't know how exactly, the message was getting across to the USAF that one of its rare aircraft was not exactly getting the five star treatment. I had been bringing photos of #122 with me or sometimes sending them on ahead to PC and showing up later. I was hanging out at *Evil I* all the time, sunny days, rainy days, snowy days, windy days, hurricanes, Nor'easters. I actually had been held over in Ohio when one of the worst nor'easters in NY history hit. I flew home on the flight from hell, listening to the compressors of my DC-9 stalling left and right.

I had USAir pilots dead heading next to me, trying to get back to New York, even they were looking pretty scared. And what did I do when I got home to NY? Ran right to the ship, water coming over the pier, to make sure #122 was okay in that horrific wind from the day before. I was recording her in all kinds of conditions. The wind on the flight deck got so bad at times, that I watched as the aircraft lifted right off the deck with only the chains holding her down by the struts. The wind was blowing up from the stern of the ship and right up her empty nacelles. It was truly an incredible sight. I could only imagine the damage that was being done to the airframe. There was also damage to me. My long, auburn hair was nothing but a mass of tangles. One time, it took a complete bottle of conditioner to sort it out and a haircut to follow because the damage was that bad.

It started to look like maybe my trips to the National Museum of the USAF (TM) might be conjuring up some good will for #122. On this particular day, while I was visiting **JV** at his office in the Research department, he got a call to come over to the PC office.
 GH wanted to talk to me. **JV** had always let **TB** another member of the PC crew and the guys know when I was going to be at the Museum, so this was going to be good. I know that **JV** and I both looked at each other. I'd never been "summoned" to PC before, so off we went to the other side of the base.

It was sort of a weird feeling walking into PC's office. While I had been there before strictly on a visiting matter, to say Hi and tell them what was going on, there had not been something this official before. **JV** and I got there, were greeted and told to sit down at the conference table in the middle of the room. One of the guys in the PC office, we will call him **SC,** dug out all the photos that I had sent to them concerning #122. The other great guy in the PC office who played a BIG part in the #122 saga, we will call **TB.** He was there also. The whole team and I sat down and the questions were fired at me. What was going on, who did they hire, what condition was the aircraft in, explain this photo, what do you know about….etc. After about 45 minutes, which seemed like 45 hours, I got the feeling that I had finally won someone's attention to take a hard look at what was going on at *Evil I.*

It was then that I was officially told to carry on. My new name would be *"DET 1".* I knew when I got on the plane to fly home that night, something had radically changed in favor of #122. I just wasn't sure what the hell I was going to do about it. I got back to New York and then proceeded to figure out how I was going to do this. It was okay that I would go to *Evil I* and scope #122 out, do some photos and send them off to PC, but somehow it wasn't going to be enough anymore. Something more was needed. I spoke with **JV** and **SC** and then **JV** went off and spoke with **SC** on the side. **JV** had come to the same conclusion that I was hitting on.

"*If you were inside the museum, you would be able to implement some changes*". So said **JV**. However, how was I going to keep it from *Evil I*, who hated the USAF with a passion. Nah, who would believe me? A Superwoman who would just walk in and save the day?!! It wasn't going to work like that. It would take more. It would take something known on the streets of New York as the "Okie Dokie". In essence, some real fancy tap dancing and evasion of the real mission. *Evil I* would never open up to me if they knew I worked for the USAF. I would have to do this completely undercover.

That next weekend, I scoped out the museum, and that continued for a few more weeks, until one day in May. May 9, 1994 to be exact. An opportunity presented itself. It was a bright, warm spring day and I was up on the flight deck. Since it was a Tuesday and I had taken the day off from work for some reason, I was up there just walking around. I saw a guy on a ladder working around #122. I'd never seen him before. He was attempting to work on some of the holes in the chine area of the fuselage. There were big gaping holes that birds were nesting in and it looked to me like he was trying to come up with some plan. Something snapped in me and I knew it was the time to act. This was a day that would live in *Evil I* History.

Chapter III-- The Spy who came in from the cold... er....? Crosstown bus.

As I was saying, there was this skinny guy working around #122. He finally climbed up the ladder and attempted to repair that damaged chine on the right wing leading edge. I'm not sure where the kick came from, but I think I know, it was a dear friend who had recently passed away by the name of George. My Ohio contact knew all about George. Later that month of May, he took a chalice to have it blessed by a very dear priest friend at the Wright Patterson Base chapel for him since he had passed away. Well, I think George is responsible for putting the burr under my saddle and getting me to walk up to this skinny guy. I went over to him, this nice, quiet, skinny guy with glasses and said" Hi, I'd be real careful about that...." He looked down from the ladder like I had three heads and answered me arrogantly " Oh yeah....Why, I answered with a smile " 'Cause its got something you shouldn't be sanding in there...asbestos" He stopped sanding and came down the ladder. I figured before he had me thrown off the flight deck, I'd better tell him my name, which I did." He told me his. From here on he will be known as *My first buddy* , who then proceeded to ask me how I knew that and continued to ask questions.

I explained to him that I was writing a book on the XB-70 Valkyrie and in my travels had come to do some research on the A-12 Blackbird for an upcoming chapter. Well, it was part true. **JV** and I had decided to write our first book on the history of the XB-70. *My first buddy* then explained to me that he knew nothing about the A-12, but his sole job at *Evil I* was to try and restore her. That scared me since he admitted that he really didn't know anything about her and really couldn't find anything to give him a lead. That was no surprise, there was nothing; no manuals ,notes, anyone to talk to. The best you could get was maybe to speak to someone who worked on the YF-12 another part of the program, but it really was a different aircraft from the A-12.

Basically, he knew nothing about Blackbirds. Well, it was a start I was looking for. I told him I'd see him around and bring him some information that he could look at. We exchanged phone numbers and as they say, the fix was in. I left the flight deck and went down to the hangar deck phones on the fan tail to call my handlers and tell them we were in.

I'm still not completely sure how I got snagged by *Evil I* but I think things just got deeper and deeper. Before I realized it, I was out there on Saturdays and Sundays. My first buddy's problem was becoming a bit clearer to me . Letters written from the *Evil I*'s aircraft maintenance staff of one, to the USAF were passed to me by **SC** in **PC** so that I got a better picture of the fact that *Evil I* was "Okie Dokie-ing" the USAF left and right. They wrote this lovely letter to USAF **PC** office telling the USAF that they had just hired a guy who's sole purpose was to take care of the A-12.

The letter described all the things that they were going to do to restore the A-12. I soon found out that the first thing that *Evil I* made *My first buddy* do when he was hired was to write this letter to the USAF! *My first buddy* told me he bulls….'d his way through it after looking at the A-12 on deck. He didn't have a clue as to HOW he was going to do this, but he was working on it. Of course, that was told in confidence since *My first buddy* didn't know that I was working for USAF **PC** at the time. Of course, I had to let **PC** and **JV** know.

So, I now knew what *Evil I* was promising, but I also knew what they were capable of, nothing. I needed some help and I needed to make this legal. I couldn't be on the sidelines feeding information anymore. If I was going to help this aircraft, I had to be on he inside. I had to "volunteer" to work at *Evil I*.

I spoke with **JV** and told him my reservations on the whole idea. I also told him that all hell would break loose if I was caught. I had to make sure that I wasn't caught. The next time out at the museum, *My first buddy* introduced me *My boss*, his boss. I was then photographed, read the rules of the volunteers and given my ID pass. I was in. I also had a real creepy feeling in the pit of my stomach and truthfully, I was scared.

Spring went into summer and I found myself out at *Evil I* every weekend and any other time I could spare from my regular day job. I had a lot of leeway there and could do most of my "road work" from my spacious office, with no questions asked.

About this time, *My first buddy* started to compile a list of things he needed at my insistence. PC had pretty well allowed me to do whatever I needed. So, now I needed a "facilitator". That person turned out to be **SC**. Now, **SC** was a nice, quiet guy, but little did anyone know that **SC** was devoted to Blackbirds and would hang it out over the line for them given any sort of provocation.

I guess I was provocation enough because I presented **SC** with a "wish list" of items for the ailing #122. He gave me phone numbers, and his blessing and he told me to tell the guys on the other of that phone number I would be contacting " the USAF will pay for it." **SC** then put the word out on his end. I started making a round of phone calls to two of my best buddies at NASA, **GP** and **MR**, Now, both these gentlemen had no fond love of the USAF given the then present situation of the Blackbird SR-71 disposition.

That along with another incident proved to make my life hell. One of my "fellow" Blackbird aficionados by the name of GJ, who I barely knew at the time, proceeded to create hell for me due to his penchant for bugging the hell out of everyone to get SR-71 or A-12 memorabilia. So much so, that he acquired an SR-71-1 manual from another great friend, LD at NASA.
GJ, after telling LD he'd never do any such thing, sold the manual to MBI (our soon to be publisher) and put it on the market. I must tell you this. LD just about lost his job, pension, etc. due to **GJ**'s contemptible actions. He also managed to make my life hell for trying to acquire materiel for #122. I'll forever "thank" **GJ** for that.

On to the story: When I reached **GP**, he just about told me to get lost. In my best, female, soft spoken, teary eyed manner, I proceeded to tell **GP** how my bird on *Evil I* was dying and she needed help. All I wanted was materiel for her, besides the USAF would sign off on it. Well, I guess that I got through to **GP**'s Montana, backwoodsman heart because the next thing I heard was "Fax me the list".

I was practically in tears of joy when I heard that and I told **GP** he'd never regret it. He warned me outright " I'd better not!" **GP**'s the only guy that I know that walked around and flew with samples of solid rocket propellant stashed in his suitcase or pockets. He was also the mentor for the Aerospike program and continues to work on scram jet engines for Gassel . But man, does he love those Montana woods to which he is now happily retired.

Slowly, packages started to arrive at Universal, my day job. Small things like 750 titanium bolts that went for $5 a pop, cans of special sealant, diagrams, cutaway drawings, special drill bits. Oh you name it, my office stated to look like the Home Depot for exotic aircraft. One by one, I started to deliver these little jewels to *My first buddy*.

I didn't want him to get suspicious as to where all this stuff was coming from, free of charge and I didn't want him to see the bills of lading, just yet. I think *My first buddy* had decided to play along and had told his boss, *My boss* all was cool. *My boss* just went along with it, too. It was obvious that I knew exactly what I was doing and knew exactly what *My first buddy* needed even before he did. So, we had the " don't ask and I won't tell" policy in place for some time. Why question the goose with the titanium bolts? I remember one incident early on in the program. I had given my list to NASA and GP and I was waiting to see if there was anything that GP could do to help us. It was a HUGE list, which included "Can" covers for the nacelles, tools, tires, lots of stuff for #122.

Actually, *My first buddy* was getting pretty damn proficient at handling her. We both had spent a lot of time with #122. And of course, as #122 told me what she needed, I passed it on to *My first buddy*. One particular weekend, I spent something like 20+ hours scraping the reverted tank sealant, better known as Viton, that had started to drip down the wheel well doors. Having no specialized tools, I made some up. Using my Swiss army knife, a nail file, along with a blunted down flat head screw driver, I *started* the tedious job of removing the sticky, gray material off the doors and any where else it had dripped and then gently washed it off with distilled water

. I ruined my nails, just about wore them down to the cuticles and dried my hands out completely since I couldn't use any hand cream due to #122's titanium . But I did manage to get the hateful stuff off of there and #122 was looking 100% better. I had just about finished this task when *My boss* came up on the flight deck to tell us the news. NBC was going to do a live broadcast from *Evil I*'s flight deck. It was a morning news show and here, it was BIG coverage. *My boss* asked us how soon could we get the A-12 ready since the producer had asked for the "sexiest aircraft on the deck". Well that was #122 and we were elated. I left work early that Monday and went down to the ship to give #122 her first real bath. I should also tell you here that another person had been added to the crew to help *My first buddy* take care of the aircraft. That was *my buddy #2. My buddy #2* and I met downstairs in the shop. The shop was the place where *My first buddy* and I hung out. This was where we figured out what to do with the aircraft who needed what etc. The place looked like a torpedo went off in there. It was often messy, smelly, oily, cluttered with aircraft , tools and paint. Instead of nudie girls pasted on the walls, there were aircraft posters. It wasn't much, but it was ours. *My first buddy* introduced *My buddy#2* to me and I knew we were going to be friends right away. We had the same sick sense of humor. You have to understand, *My first buddy* was a sweetheart and we loved him but *My first buddy* was your true New Englander. He initially was from Vermont and had transplanted himself in New Jersey. Both *My buddy#2* and I were from Queens and both of us had street smarts, not to mention street attitudes. All of us went down to the shop to figure out what to do for this program and how to get #122 ready in time. Okay, I was just about to give her a bath. *My first buddy* had a couple of repairs to make and *My buddy#2* was busy supporting both of us. I waited till both the guys took their stuff on deck and I was about to make a phone call to GP at Edwards to see if he had any luck with the list. I knew it was going to be too much to ask. I was just dialing the phone when *My first buddy and My Buddy #2* flew into the shop completely wild, bug eyed wild. I didn't know what the hell had happened. It turned out that a big, flat bed truck had just pulled up to the gate with 4 pallets of materials on it. *My buddy #2* handed me that bill of lading. NASA/Edwards was all over it and so was my name. The guys grabbed stuff to open the pallets and went back down to the pier to supervise the unloading of the truck. I just sort of smiled to myself and said a mental thank you to **GP**. If ever anyone saved the day, it was him and his timing couldn't have been better. *My first buddy* and *My buddy#2* looked like two kids at Christmas. I was on my way up to the flight deck and I was passing *My boss's* office. I just smiled as I went by and *My boss* just stared at me, like he didn't know what to make of me. Actually it left me a little uncomfortable maybe because I knew what was really going on and he didn't. When I got outside, there was My *buddy#2* and *My first buddy* tearing crates open like two wild men. **GP** had granted the whole list. Can covers, tires, brand new silver bolts, RTV sealant, just about anything and everything I asked for.

My boss and *My first buddy* had stood there amidst it all and just suddenly stared at me. On that note, I went upstairs to wash #122, quick, before I had to answer any questions. **GP**, NASA and **SC** had my undying devotion that day. #122 would look fabulous for her TV debut. I washed her by myself, all 101.6 feet of her, on my hands and knees. When I got through, she looked like a million bucks, or more like the 8 1/2 million it took to build her. *My buddy#2* and *My first buddy* brought up the can covers (which by the way, were the only surviving piece of equipment from 934, the ill fated YF-12) the covers even had a wonderful bat/snake logo on it *My buddy#2* and *My first buddy* lifted the can covers into the empty nacelles and suddenly #122 *started* to look "hot". The guys pumped up the ties for now and I scrubbed the side walls to shine them up a bit. #122 was looking great! She was going to look great on TV too. Of course, I made sure I let **JV** and everyone else know about the show.

I gave *My boss* the finer points of A-12 history and I have to say when I turned on the TV the next morning at 6:00am (yes, the VCR was up and running) #122 looked astoundingly gorgeous, so did *My first buddy*, *My buddy#2* and *My boss* and some of the other *Evil I* crew that was on the deck for the show. #122 was the backdrop for the entire show. I was so proud and pleased that the stuff arrived just in the nick of time for her debut. She really looked like a class act.

Chapter IV: Blackbirds and "procurement" (sung to the tune of Pirates of the Carribbean-Curse of the Black Pearl ™)

To quote Jimmy Buffet---"Yes I am a pirate" and my traits in"horsetrading" came in handy.

To say the least, my first shot at finding stuff for the A-12 was a quantified success. However, I knew from then on that I was going to have to watch things closely to make sure that no one got suspicious as to how and why I was able to get the strange items I needed. There were "eyes and ears" on *Evil I* that were not happy with the new success of the aircraft department. The strange thing is, this new talent was also being called upon to help the other airplanes on the deck. Not that I ever for a moment minded that. I was more than happy to beg, borrow or appropriate anything I could to help those aircraft who were is such desperate need. The list was as long as my arm: *Vought F8 Crusader, Vought A-7 Corsair, Grumman F-14D Tomcat, Grumman EA-6A Intruder and the EA-6B Intruder II, Grumman Albatross, Grumman E1B Tracker, the McDonnell Douglas A-4 SkyHawk, McDonnell Douglas AV-8C Harrier, Bell AH-1J Cobra helicopter, Bell UH-1M helicopter, Douglas F3 Skynight, North American FJ-3 Fury, North American RA-5C Vigilante, General Dynamics F-16A.* They were quite a handful!

As long as *My first buddy and My Buddy#2* were willing to do the work, I was able to find the stuff.

#122 was really starting to shape up nicely. She was getting a bath just about every month as the weather permitted. We were all discussing her over all condition and what to do next. It was decided that *My first buddy and My Buddy#2* would initially start some repair work on the chines, (that is the curved edge on the fuselage) and of course, there were the pie shaped panels that were underneath the elevons on the aft section of the aircraft) that had to be removed and new ones cut out of sheet metal. We had to get the asbestos out of there. *My first buddy and My Buddy* #2 and I cut each panel by hand to match the panel they took out. That was something like 118 panels that needed to be changed out. The important thing is that she was getting aggressively fixed. The words and photos were being passed to the PC office and to JV. PC was actually happy. Yet still, I had to work under cover. I never made a phone call to the Museum from the hangar deck or from the shop for fear of it being traced.

There were a lot of nosey, jealous people on *Evil I* who weren't very pleased with me, or what I was doing. Sort of a turf war had broken out between the Operations department and the newly developed Aircraft curator/Maintenance staff. *Museum Director,* the director of the museum if you hadn't guessed, was elated with the current situation and basically protected us from the wolves.

Not only was #122 starting to look good, but the other aircraft were getting some badly needed attention, too. *Museum Director* was getting the kudos from all the board members that the flight deck no longer looked like an eyesore. People were starting to notice. Since SC was so wonderful at helping me out, I decided to expand operations to the Navy.

I met another great guy by the name of **Navy Z**. **Navy Z** as we will call him to protect him, was responsible for all the Navy aircraft at *Evil I*. He worked out of the Philly Naval office. **Navy Z** had been praying for the day that someone would take the aircraft on *Evil I* seriously. He had been complaining for years about the poor condition of some very important Navy jets on the deck. Well, the aircraft department was now there to assist, but DAMN what a mess!!! I personally could only look at it one airplane at a time because anything else would have sent me screaming off the fantail. You must consider that many of these aircraft had been out on deck all year long, with snow, ice, rain and no treatment at all. The stuff we found would fill a book in itself. So to say the least, **Navy Z** was happy to work with us. The boys were giving me the lists and I started looking for stuff. I was really amazed how I managed to locate stuff like cockpit canopies for our Skyraider and Neptune and Vigilante, and parts for our other residents.

I knew I was learning the ropes pretty well on how the Navy worked. One thing I did find out was that *Evil I* had a pretty poor reputation as far as being a museum was concerned. Every other museum I contacted was fairly in shock to find we were actually getting our act together. I heard a myriad of terms put to us like " Floating casino, 12th Ave Ballroom, Bar Mitzvah palace".

It really did bother me. While I knew that museums did use some events to pay the bills, it was looking to me like *Evil I* was just that, a floating ballroom. More attention was paid to the hangar deck and the condition of the rug than to the exhibits. The Events that were being held were damaging and highly questionable, like the New Year's Eve Ball that left some of the aircraft vomited on, glasses of booze left in the cockpits, other pieces of disgusting personal paraphernalia and wing damage. Sadly enough, I couldn't counter the stories, but I made it my business to tell *Museum Director* and *My boss* just what people thought of us. While maintaining my undercover status, I worked both systems Navy and USAF, made the contacts, got past the bad attitudes of not only the people I was asking for help from, but the rather nasty attitudes of the Operations department on the *Evil I*.

They were extremely bent out of shape by the fact our little aircraft office was getting the flight deck together after years of their neglect. A certain select few, by the names of *Ace*, *Buddy*, and *Lou* (who we later converted to our side of the ship) were hell bent on making us fall flat on our face. You see, the flight deck had been their private my domain, much like the rest of the ship. *Ace* lived and ruled like king of the roost ,which would be more appropriate. He made our lives hell any chance he could.

To be blunt, "Ops" was the parasite that bored itself a nice little hole in *Evil I*'s administrative guts. They were hanging in there like any other good little tape worm.

We, in the aircraft department, went on our way and were challenged at almost every step by Ops. The *Museum Director* finally had to step in and tell them to back off, but the damage was already done. It was a pissing contest every time we had to do something on the flight deck.

Little did we know then that one of the reasons Ops had a problem with us was because *Ace* was collecting and protecting his own little stash of misappropriated government surplus goodies and we were getting too close to finding just what he was hiding. While I was going along and getting the things that were needed to work like a real restoration department regardless of the Ops Department, *My first buddy and My Buddy #2* were already running into problems on the flight deck.

We were often limited by factors like, bad weather, heat, humidity or freezing cold along with the high winds that were coming off the river. So we tried to get as much in as we could on the good days.

I had begun making contacts with government agencies like AMARC (Aircraft Maintenance and Regeneration at Davis Monthan AB) and GSA (General Services Administration) to get some of the surplus stuff we needed. One very special aircraft to me was the RA -5C Vigilante that we had on deck. I vowed to myself that somehow right after #122 , she would be the next to get some desperately needed restoration.

The RA-5C Vigilante was the baby sister to the XB-70 Valkyrie. She was the Navy's first Mach 2 aircraft to go off a flight deck. The "Viggie" as she was known as a nuclear strike bomber or Reconnaissance aircraft, made by North American Aviation,

The RC-5A Vigilante on the deck of the USS Enterprise
(US Naval Aviation news, circa 1962)

I really wanted to make a show piece of her. I was also beginning to see in myself that I was getting in deeper than I ever had thought possible. I was no longer just working for #122 and the USAF, I was now the den mother to the 28 other aircraft that always greeted me when I stepped on deck. I was beginning to just feel torn in half. I don't know why now and I sure didn't know then. I guess I was just accepting it as part of my cover. I could feel that indescribable something pulling me in closer to *Evil I*. I wasn't sure if I couldn't or didn't want to get out.

As I was washing #122 down one day after a really bad rain storm the day before, mostly cleaning up the grungy struts, *My boss* came on deck and asked me if I could work on a special project. I asked him what was up. Well, he said we had just acquired two new aircraft. Well, I had heard about that from My *buddy#2* and *My first buddy*, but only that it was in the working stages. *My boss* explained about the F-14 and the EA-6B Intruder we had just acquired. It was a gift from Grumman aircraft on the Island, Long Island that is.

They were closing their Calverton facility in Riverhead, Long Island. (a REAL piece of history) and these were the last two aircraft left. Grumman was a long time buddy of the *Evil I*.

They had donated aircraft to us before like the First E-6A Intruder off the line. She was in our possession, but had a couple of problems including a crazed nose. It could be repaired, if I could find the nose. (I was already working on that with Grumman) Which is how I met a great guy named **JR. JR** was in charge of the EA-6 Weapons systems program for Grumman. He was a "real paison." He was just a great gentleman, rough diamond with a heart of gold. He had to be tough. He was always in Washington fighting for his program. Well, you guessed it, I said yes to *My boss*. I knew *My first buddy* was already working out the details of the move, so yeah, I was in.

It took about another 5 or 6 weeks before moving day, so I was still busy trying to get in for our various other problem children. I swear, the flight deck was actually starting to look like something. *My buddy#2* and *My first buddy* had a new project just about every day going on up there. We were still working on #122 and I was still letting **JV** and **PC** know what was happening. PC had kept their promise of zero involvement, in essence, kept the pressure off *Evil I*. So, *Evil I* actually thought they had sent the USAF running with *My first buddy's* previous letter of intent. *My first buddy*, in the meantime was running back and forth to Long Island, measuring, riding the road and doing what every other good "trail boss' (that's what we were calling him) would do in a move like this one. It was no easy job. He went out and measure how many feet high the railroad trestle was on the route, how many inches the curbs were, how wide was the roadway. It was incredible. *My first buddy* was totally amazing. The move was getting closer, which as I remember it was in mid April. We'd all gone over the plans, *My first buddy and My buddy #2* killed themselves making sure we had everything we needed, tow bars, chains, air pump etc. The concept was to walk the F-14C and the EA-6B from the Calverton hangar through the town of Riverhead and out to the docks on the other side of town to the waiting barges where they would be brought up the river by the by *Evil I*'s tug boats to the *Evil I*.

At least I wasn't too worried about these two guys , they were made to travel in water. Of course, there was the usual interference by Ops, because *Ace* was driving the tugs and barges and had everything to say about how *My first buddy* was doing it all wrong. *Museum Director* told him to shut up and stand by, which *Ace* did very grudgingly. The day had come, or I should say the night. We met at the ship at 11:00PM the night of April 19th. When *My buddy #2* and I got there we saw an often familiar sight, two police divers jumping off the end of our pier looking for a supposed suicide. While we waiting for the rest of the team to assemble, we messed around and drank lots of coffee.

Of course, Ops was there to wish us ill and *Ace* pulled *My first buddy* over and was talking to him like he was some little school kid or something until finally the guys from the USS *America* showed up and we were out of there, gratefully, leaving *Ace* and the rest of Ops cursing us in the April breeze. I have to say that I appreciated the fact that *Museum Director* had faith in *MyBuddy #2* and *My first buddy* and told Ops to basically get stuffed and wait for the radio call that we were near the docks, so that they could meet us. Of course, as it turned out OPS was LATE and had every excuse in the book including the tide was against them, but you'll see that later.

By 11:30pm we had climbed into our cars and trucks and joined by the *America* crew , along with our own guys. We left *Evil I* in a caravan. I have to say this, I was the ONLY woman on this road trip and I was treated just like one of the guys (which I accepted as a compliment) I was spared NOTHING!

Which translates to one of the most hysterically funny rides down the Long Island Expressway for two hours that I have ever been involved in. It was rude, crude and very funny. We finally arrive at the Calverton facility somewhere about 1:00am .

The move had to be done at night because the town didn't want its traffic messed up in morning rush hour. When we pulled into the Grumman facility, *My first buddy* checked to make sure we had all the permits. We stopped at the guard gate to be checked in. You could see and feel the history all round you. While it was dark, spread out and desolate, the ghosts of Grumman's fantastic history was there. We passed security and pulled up to the hangar. The doors were open and the inside was flooded with light. Bathed in that light were two beautiful aircraft sitting side by side. You know, I choked up when I got out of the car. The scene is still burned in my mind.

The warm lights of the hangar were glinting on the white skins of these two aircraft. With the two aircraft were their crews, the last hold outs. These guys had taken care of these aircraft all their working lives. They had helped to build them. Everyone was quiet because all of us knew how these two aircraft would be the last to leave the facility and then it would be closed for good shortly after their departure. *My first buddy and My Buddy#2* went over with the *America* crew to speak to the guys from Grumman and get their plans on the same page. I walked over and touched the F-14D. As always, I was moved by the intensity I felt as I ran my hand along the fuselage. Power, raw and proud. I told her not to worry, we'd take care of her.

I then went over to the EA-6B, and I felt the same raw, proud power. I fell in instant love with both of these warriors, and I felt the love and soul of Grumman pour out of them. It must have showed in my face. I saw *My first buddy* watching me from the side. He walked over and touched my arm. He said " *Tell them it's gonna be okay*".

At that point I knew that *My first buddy, My buddy#2* and I had bonded. *My first buddy and My Buddy#2* both knew I talked to airplanes. Maybe they didn't totally understand, but they believed. Somehow, I always had the feeling that was the reason that I was brought along that night, to let the aircraft know it was going to be okay. We were just about ready to go when I saw the last of the Grumman crew bid a farewell to their aircraft.

Everyone was choking up and I was standing there with tears in my eyes, much like right now as I write this. I watched the *America* guys salute the Grumman guys and thank them for their service. The *America* was an aircraft carrier and her guys still flew the EA-6B and the F-14D. There is no way to put on paper the emotion charged atmosphere.

It was so silent in there for those few seconds until the emotional transfer was made. The aircraft now belonged to us. Everyone started breathing again, as the guys started up the trucks and started pulling the aircraft out of the hangar. Now the fun started. Once we left the Grumman facility, we would have to walk the girls twenty two miles to the wharf where we would board them on the waiting barges.

It was now almost 3:30 AM, and it was going to be a long morning. We rode right behind the two aircraft as they went out of the gates. When we stopped just out side the main road, it was very dark and chilly as I got out of the truck to stand beside the right wing of the F-14D. I would stay there, next to her, through the long march to the docks which was twenty two miles. The walk was not uneventful, though it was a slow procession through the streets of Riverhead.

While *My first buddy* had planned for every bloody contingency, there were some unknowns that cropped up. The EA-6B was up ahead of the F-14D . *My buddy#2* and half of the *America* team and some of the *Evil I* people were with the EA-6B. *My first buddy* was between both aircraft and the rest of us pulling up the rear with the F-14D. .

Our first encounter with a problem was with the train trestle on the way into town. While *My first buddy* had measured every inch, the Intruder wasn't going to clear it being just about 2 inches too high. The *America* guys and *My first buddy* decided to let the air out of the tires to clear the EA-6B and then pump them up on the other side. It worked like a charm. The F-14D was luckily already low enough to make it. The sun was just coming up as we reached the outskirts of the town of Riverhead.

I had been doing a slow jog next to the F-14D and was grateful when we stopped for a couple of minutes. We had just turned into the main road, you know the type, two lanes coming and going with a meridian down the center, strip malls on both sides. I was amazed to see all the townspeople out there!!!! I think we were shocked. Grumman had been a huge part of Riverhead since its inception, the town had actually been built around the first plant. So, we were greeted by applause, offers of coffee and doughnuts, like marathon runners. They even had the local traffic helicopter over head doing color for the radio.

There were people doing videos, some taking photos, news photographers, the whole shebang. I did see a photo of us in the Daily News the next morning, and I heard that we were on the morning news on channel 7. We were moving slowly so everyone got a good look. The response of the town actually made it a little easier for all of us. I think the girls enjoyed it ,too. I was getting tired so I was holding onto the wing tip of the F-14D as we were walking slowly. It really felt good to lean on someone with that kind of strength. Looking ahead as we slowly moved up the street, I saw the sign posts. There was sign post coming up and I realized with a shock that the F-14D , even with her wings pinned back, was NOT going to clear that post. I started a run quickly and just made it to the sign as she was moving up. I took a flying leap onto the sign and attempted to pull it back so she wouldn't clip it with her wing tip, which would have really messed up things for the F-14D and for the town. My small stature wasn't going to be enough to hold it. In my loudest voice, I yelled " HOLD IT!!". It was amazing to see the caravan stop so quick on that one call, but they did. The F-14's wing was literally centimeters from clipping the sign, with me trying to hold it back so it wouldn't snap and hit her. I really wasn't strong enough to hold it much further and my fingers were just short of being stuck between the sign and her wing.. Thank God I saw *My first buddy* and one of America's crew running up to me. *My first buddy* jumped up on the post and pulled it back further while *America's* crewman got me down from the post. *My buddy #2* signaled for the F -14 to be backed up a little so we could get some clearance. *My buddy #2* got her cleared. I sat down on the grass as I needed to catch my breath. *My buddy#2* just looked at me and laughed. He said he couldn't believe that he heard me all the way up by the EA-6. He said everyone just stopped cold up there. *America's* crewman came over and said that even though I was small, I had a helluva voice, thank God. *My boss* had run down from the EA-6 to the scene. *Comment :" What the hell did you do now?"* (I was noted for getting into tight squeezes on the ship)to which *My first buddy* said, " Kept the damn wing from getting ripped off."

My first buddy and the *America* Crewman made the decision to pull the post out. The local cops came down to see what happened and they agreed that as long as we put it back up right away, there would be no problem. So, a team was made to pull sign posts and put them back after the airplanes passed. It slowed things down just a bit , but better safe than sorry.

We finally made it out of the town center and headed for the residential area and the docks. It was already getting late. This had taken much longer than we had anticipated. By 1:00PM we were almost to the docks, using the residential roads. People would stare out their windows or come out on their porches because they couldn't believe the sight that had pulled up on their front lawn.

Speaking of lawns, we did mess up on guy's curb. *My boss* went straight to the front door, knocked and when the occupant came out he accused us of being with Candid Camera!. It took almost fifteen minutes to convince this fellow that we were legit, and all we wanted to do was pay for the damage to his lawn. It turned out that this guy was a real gentleman and waived the damage off He just asked would it be okay if we took his picture with the airplane. So, it took another half hour before we could get on our way again. It was a small price to pay and our insurance company would be happy. The best and we all agreed, the funniest thing on this trip had to be the four-year old little boy, who was playing on his lawn. He ran like the devil inside the house, when we showed up to tell his Daddy the airplanes were here!! We could hear him yelling from inside the house for Dad to come and see.

Finally, the guy did look out the window and again, the face was priceless. The Daddy came out and stared, completely in awe. All of us broke up when the little kid said " See, I told you!" Well after another ten minutes of Video and questions. We were finally able to get on our way. We were just about twenty minutes from the dock. It only took us another two hours to get there. As we approached the dock, we were encountering a lot of soft ground.

The Intruder slipped off the edge of the road and into the soft dirt. We all scrambled around looking for something to slip under the wheel to get some traction. Although *My first buddy* had planned for that contingency, the wood we had wasn't quite enough. One of the guys finally found a cardboard box in the back of one of the trucks. Have you ever tried to "rock" an A-6 Intruder? Believe it or not , *America's* crewmen did and got her out of the hole. It did take us about an hour to get that done.

We finally did reach the docks only to find out that the barges had not yet arrived. This meant that we would have to wait with the aircraft till the barges did show up. Since we had been up just about 24 hours, 90% of us had not slept since the day before, me being one of then. We figured to get some food and relax a bit. We also took the "end of the line" photo with the F-14D and the EA-6B. Each team, with their airplane. I went off with *My buddy #2* to find somewhere to get sandwiches. We found a roadside deli, who loaded up a tray with sandwiches and coffee and soda and off we went. Everyone was finally unwinding and we contacted the barges by radio. They were just about 2 hours away. The *America* crew decided to wait, and escort the aircraft back on the barges. I was offered the ride but quickly declined. My "allergy " to boats quickly became the butt of many jokes. That was okay, I didn't mind as long as I didn't have to get on the boat.

My first buddy was going to stay with the aircraft and the rest of us were to head back to *Evil I*. So I bid the girls farewell and headed back home. I was never happier to see my bed, since I was sore, achy, and very tired, but at least I knew our "new girls" were safe. The aircraft actually arrived safely the next day. Fortunately, I was not there when they hoisted them on deck via the crane from the barges. I probably would have had a heart attack. I saw the photos late it was better that I wasn't there.

Materials for aircraft repair were now rolling in from various places all over the country. The word was getting out that *Evil I* actually looked serious about becoming a museum. Even the Navy Curator's office took note of our new methods. We had actually moved two aircraft safely and got them back to the ship. We were proud and OPS was pissed we had done so well.

Chapter V: the "*Evil I* Traps the Spy."

Procurement wasn't the only thing I was getting involved in. I found myself being dragged down even further into the world of *Evil I*. I was finding that things were being thrown my way. Things like being there almost every weekend or days off. Day and night my phone was ringing about airplane problems on deck, etc. I woke up one morning and realized that I had two full time jobs. Not only was I serving the USAF and taking care of the A-12, I found myself deeper into the *Evil I*'s problems. They were trusting me fully and only I KNEW the other half of the coin. If they ever found out.... I didn't know who I was working for anymore, *Evil I* or the USAF. I was trying to maintain my relationship with the USAF without the *Evil I* knowing it. I often found myself on the edge of a dilemma, while things were getting better at the aircraft shop and for the airplanes, things were getting tougher for me. I felt split up the middle. Did I have allegiance to the USAF, and loyalty to the *Evil ?* Yeah, right. The *Evil I* was definitely worried about the USAF. I finally got to see the letters that were sent to them by PC, some of then were really threatening. *Evil I* had felt that the USAF was pushing them. The real problem though was *Evil I* didn't heed the contract they had signed. In fact, **Navy Z** told me that the Navy would give the *Evil I* nothing more in the line of aircraft until they got their act together.

Evil I or should I say "OPS" department who was taking care of the flight deck back then basically held everyone in contempt, including the loaning of the aircraft. So when the "new" aircraft department came into play, things were taken away from "OPS". You knew they were getting nasty and nervous. It was also showing at the Museum. OPS attempted at every turn to make life miserable for us. They usually succeeded. Nothing we did was right.

They were constantly harassing *My first buddy* and *My buddy #2* and the same went for me. Always looking to start some trouble, they wore a hole in the rug to *Museum Director*'s office. It wasn't until *Museum Director* truly made us official that the "OPS" department was forced to back off and let us alone.

Things were getting tougher for me because I couldn't let *Evil I* know I was working for the USAF and meanwhile while *Evil I* trashed the USAF, I had to join in for fear of some one noticing I wasn't on the team. It took me quite a while but I finally convinced *My first buddy* of the seriousness of the contract situation and the only way you would get rid of the USAF and the forever threatening letters was to adhere to the contract.

Too many people on the *Evil I*, right down to the former crew members thought that *Evil I's* owner could fix anything., so nothing was sacred. They could get away with anything, if the owner threw money at it. That worked on occasion but it was destroying the credibility of the Museum. I convinced *My first buddy* that the contracts could be called at any time. In short, the Army, Navy, Marine Corp and the USAF could inspect at any time and if they didn't like what they were seeing, they could pull an aircraft or make it very ugly for us. I know *My first buddy* tried to get the message through to *My boss*. I got dragged in to explain to *My boss* just why I was so nervous about our contractual situation.

Now, you've got to know that PC and JV knew exactly what I was doing. *My boss* did speak with *Museum Director* about it. I could also see OPS getting nervous. They had always overseen the contracts and as we found out later, were making their own deals, many of them shady. So, it was no big wonder why they were in a panic and demanding to know WHY the Aircraft department wanted all the paper work. *Museum Director* finally agreed that the contracts needed to be put into some sort of order. While not exactly admitting ignorance about OPS and their deals, *Museum Director* knew it had to be done. He was enjoying the kudos from all sides due to fact the museum was started to act like one. *My first buddy* and I were told to get the job done.

When OPS was told to pull together paperwork they were hoarding since the museum opened in 1977, that's when the proverbial shit hit the fan. There were protests, threats against us, until *Museum Director* warned them no more shenanigans or they'd be looking for new jobs. Well, *Ace* and his crew didn't take too kindly to that. However, they started to hand us what they "wanted" us to have.

I was at the point in this relationship that I was started to stress out. It was getting hard to try and stay objective and above it all. There was no way out, so all I could do was try to straighten the mess out.

Between, *My first buddy, My buddy #2* and myself, we went to work. It was going to be a long, arduous project. I was still fighting with myself on the loyalty point.
I wanted to sort out #122's mess, but now I was sorting out *Evil I*'s mess too. I shut up and dealt with it. It didn't matter now that my health was starting to get affected and I was beginning to feel drained.

My first buddy and I started to collect the contracts and sort out who came from where. It was amazing to see that so much of the stuff we had, was not under any contract, it had just showed up at the Museum!
That was the way it was done when the Museum opened way back when. We also were starting to find out that there were things missing. We now had to go and check each piece of my list; check the tail number, the serial number plate etc., on all the aircraft. That meant climbing in, out up and down and over aircraft. I got all the small, cramped spots, bet you can't guess why.

We found out that we were missing at TBM. Yes, we had the girl that was on the hangar deck sure enough, but apparently there was another one somewhere around 1982. We asked OPS and *Ace* who has been with the Museum since it opened, much like *Museum Director*.

No one remembered anything. We were to find out lots of people around the museum had short memories or none at all when it came to certain things. In any event the process had been *started*, we were certainly ruffling lots of feathers. While we were doing the aircraft, I thought it would be a good idea to also check the missiles out.

We had them on display all over. The Phoenix missile that was with the F-14, Sparrows, Shrikes, an old Hound Dog, Stingers etc were cataloged with everything we else we had . I also checked to see if they had been de-milled (demilitarized), since we had no paperwork as to exactly WHERE these missiles came from.

 Once again, no one really remembered. I asked the Navy for help. MW, the Naval Historian/ Curator was a dream. He helped me find out how to look for fuses and how to open warheads, etc. Something every girl needs to know in this day and age. Most of them were okay, but I did find a couple that were questionable. I was told to pack them in their cases and ship them to Point Mugu in California. **Big Z** had a Navy truck pick them up from us, as they weren't going to travel via Fed EX. About a week later, I got a call from Pt. Mugu's Ordnance Officer.

I got the call at work at Universal, so it was real interesting when call was announced over the intercom " Naval Ordnance is holding". I remember the look the owner ML's face. Priceless!! Pt. Mugu told us that while these were basically inert, they needed to be EOD'd which basically meant completely disarmed and certified as such.

Well, I called *My boss* and told him the story. He told me to go ahead, what ever it cost, better safe than sorry. So, I called Pt . Mugu and told them to do it. We got the missiles back in about 3 weeks; cleaned, EOD'd and certified. There would come a day when I was GLAD I had done that, along with having a precise inventory of every missile on the property.

There was a great thing going on throughout the country that spring. It was called **Freedom Flight.** This was a flight comprised of WWII AT-6's, B-17's and many of the other warbirds that were being flown privately or by Museums. It started out on the west coast and was working its way eastward. Matter of fact, JV had the **Freedom Flight** over in Ohio. Well, they put *Evil I* on the list as fly past point. *Evil I* was pleased to make it a party for all. The flight deck needed to be in order because we were going to have a massive VIP list, USAF chief of staff, CNO of the Navy, Francis Gabreski WWII ace, lots of VIPs, the Mayor and a cast of hundreds. It was by special invitation to the flight deck and security was monumentally tight. It didn't make any difference to us, we had to be there. It was our deck.

Okay, where to start. Since the party was going to be on the stern of the flight deck, the Neptune, the BIG friggin' Neptune needed to be spruced up along with the Albatross (who was in poor ass shape), we had to move the MiG (not fitting), haul the F-4 Phantom over and wash her, too, and oh yeah, could you do something with the helicopters? Yeah. I could do something. The E-2B Tracker needed to re- moved to make more room but make sure she looked good. OPS was screaming because they needed to put up the tent and we were in the way. In short, it was basic chaos.
Museum Director was nervous as I'd ever seen him. We were all exhausted. *My first buddy* was painting the Neptune's props at 1:00AM in the morning by flashlight (the deck lights weren't enough) and I still had to clean up the F-11 on the pedestal.

The other half of the flight deck, including #122 had to be cleaned. This went on for about a week before the event. I was at work when I got a phone call from *My first buddy. He* said,. " *Any chance you could find us a B-17 for the Freedom Flight?" They don't have one in the area and Museum Director and the board want one."* So after I finished choking. I asked him where did he think I'd find one.

" You always do." Yeah. So, I dragged out my trusty phone book and started to think of who I could call. Jeff Ethel, who has since "gone west" God rest him, was the phone number I fell on. ' I called him and asked the impossible. I though maybe I could get his girl *Fuddy Duddy* from Genesco in upstate New York. Jeff told me that she was in maintenance and beside, there wasn't enough funding for it, but... How about Bob Morgan? Bob Morgan was the pilot of the Memphis Belle, the REAL Memphis Belle. He is also my wing in the CAF(then Confederate Air Force). I asked Jeff if he knew where he was. Jeff said he was in Philadelphia at a show. Jeff offered to call him and have him contact me. Jeff, I know you 're up there floating on clouds, forever, you do have my undying love. About a half hour later, I get a call at work. It's Bob Morgan wanting details on how to hook up with **Freedom Flight**. So, production went to hell that morning at Universal while I set up getting the B-17- Sentimental Journey ,for the *Evil I*. It was all arranged in just about an hour. *Evil I* would pay for Bob's fuel to fly up from Philly and meet up with the **Freedom Flight** crowd. *Evil I* would be contacted by radio when he reached the George Washington Bridge. Museum Director was forever grateful. I owed my ass to the aviation world.

By the morning of the event , 6:00am to be precise, *My buddy #2* and I dragged our exhausted butts up to the flight deck.

Museum Director wanted everything inspected…everything! Okay, *My buddy#2* and I started to inspect….everything. I saw that somehow some dirt had gotten onto the Corsair on the #122 side of the deck. I figured I could just rinse her down. *Museum Director* was right behind me. " Oh shit look at that!! Jeannette, get in there and clean it up." It was now 7:30 AM. The guests would start arriving by 9:00AM and I still had 3 airplanes to check, and I still had to tie the Helicopter blades back with *My buddy #2*.

There was no sense trying to explain this. I went down got a bucket of water and sponges and rags and some of the stuff we used to wash down the other aircraft. Thank God, I wasn't dressed yet, this was going to be a dress up affair, so I brought my good clothes. I sat my sorry butt down on the deck and *started* to clean this dirt from hell off the Corsair and the scoop she has in front.(I still swear that OPS planted that dirt) So here I am at 8:35 am, sweating, my hair now looks like shit, I chipped my nail polish and I'm just about ready to kill *Museum Director*. My buddy #2 and *My first buddy* were on the other end to the deck fighting with the F-4, when here comes *Museum Director* with the Chief of Naval Operations, Admiral **JJ.**

He was giving him a tour of the Museum. Since I was staring into the scoop and washing it out, I didn't know that my butt was saluting the CNO. *Museum Director* raised his voice so that I would at least hear them behind me and I pulled out and turned around. Dirt on my face, wet, my jeans soaked, I turned around to greet the CNO. *Museum Director* says. " Admiral, this is Jeannette she's part of the flight deck crew and the aircraft historian for the *Evil I*.
(first time I heard this!) My *buddy #2* was standing behind the Admiral and *Museum Director* and almost choked from laughter. I would kill him later. All I could muster was " Good morning, sir." He smiled and nodded " Nice job. She looks great" Thank you, sir". *Museum Director* was now beaming. " Well, I'll let you finish, hurry up, Jeannette. You need to get changed." I smiled and watched them walk away and head down the main staircase at which point I picked up the bucket, but *My buddy#2* was too fast and grabbed it before I could do any damage.

The day of the show was gorgeous, it was warm and sort of sunny, there were some clouds and out by the George Washington Bridge, it was a bit hazy. I had just enough time to repair the damages to hair and nails and change clothes.

The VIPs and the airplanes were arriving as scheduled. The barbecue pits were open on the flight deck and gearing up to start serving lunch. I saw the USAF Chief of Staff, General RF with his aide in tow, going over to Francis Barresi's table right on the edge of the deck, which was fenced in and covered by the tent
. Pictures were being taken left and right. Basically our job on the flight deck was for us to mingle and help the VIPs' when they needed it. Mayor RG was there and gave a short speech, he then sat down with *Museum Director* and a couple of the other board members and they started gabbing. *My first and second buddies*, were basically clearing tables, finding seats and locating people, making sure everyone was having a good time. There was a ton of security on the deck. There were ceremonies with wreaths being tossed off the flight deck into the water below commemorating Pearl harbor, *Evil I* Crew, Navy and USAF flyers from WWII as the first of the AT-6's rolled in and dipped low just in front of *Evil I*. I have to say it was both moving and very exciting. However, I was waiting for one special girl, *Sentimental Journey*. I must have gotten asked 100 times when is she coming.

According to Bob, he would be there just about 12:30. He told me forget the radio and look out towards the bridge, I'd know when he was there. I kept looking and sure enough, out of the clouds just like out of a dream, I saw a B-17 coming. The weird thing was, I pointed to the haze out by the bridge and *My buddy#2* was looking but he didn't see her.

I did, I always knew when an airplane was coming because I would look up before she got there, I always knew when they were near. Then the sound of those magnificent engines came roaring towards us. *My buddy#2* was yelling for *My first buddy* to get on deck, *My boss* was hanging over the side to get a better view and I could see USAF Chief, Gabreski, Navy CNO, Museum Director and the rest hanging on the fence at the end of the deck. Like some big, beautiful angel, *Sentimental Journey* slowed to pass *Evil I* and drop rose petals in the water in front of us to join the wreaths in the water. The USAF Chief saluted along with the CNO and Gabreski, I was in tears as usual and *My boss,* and *My buddy* #2 watched in awe as did the rest of the flight deck. Everything just stopped as she passed. It was a sight I would forever treasure, it was a once in a lifetime thing.

After grabbing some of the napkins off the table to dry my eyes, *My boss* told us to relax a little bit. He started piling up the plates with chicken, hot dogs and hamburgers. M*y boss* had a table off to the side of the main area, he had already had a couple of Heinekens once the word was given that the crew could stand down. He started piling the plates of food over there for us. You didn't have to ask *My buddies* twice and even I was hungry at that point. *My boss* was as relaxed as I had ever seen him. He put his feet up on an extra chair and looked out at the rest of the aircraft on their way past *Evil I*. He sipped his beer. I will give you verbatim what he said. It will never leave me.

" This is good. This is a great day...... You guys made it happen... ... You put up with so much and got it all done.... My buddy #2 as always, pitching in doing the worst jobs and making it great.... And you Jeannette.... Always there, when ever we need something, just say the word.... two ...three days later...POOF! There it is...ask for a B-17 Poof! There she is....but you.. You're an enigma. I don't know where you came from...I don't know much about you other than what goes on here... You never say too much, you never say too little.. Always just the right amount for the situation.. You never give it away. I don't know how you do what you do... And I don't want to know. Just keep doing it."

I was totally stunned by this. He tipped his beer to us and then turned to look out at a pack of P-51's passing by. *My buddies and I* looked at each other. We'd NEVER heard anything like that come out of *My boss*. *My buddy #2* offered that it must be the beer and of course, *Museum Director* was getting the praise and was terrified, maybe I was too obvious... I didn't know what to think other than both *my buddies* told me that *My boss* never said much but when he did, it had to be important.

My *buddy#2* drove me home from the ship after the party . I was still bothered by what *My boss* said. It was making me feel worse because I still had the USAF on my mind. Talk about getting torn up the middle. *My buddy#2* was even thrown by *My boss's* little speech, but not to worry, we'd all be swearing each other out by the weekend he assured me

Chapter VI Blackbirds, grunge work and dancin' on the edge.

Things were progressing on #122. I had talked with GP about lending us the special jack used to change Blackbird tires. As always, not to worry, the jack arrived almost a week later. *My buddy #2* had taken the tires to an auto shop and had them filled with foam-fill. This meant that she would never have flat tires again. They also were brand new and shiny silver which would make #122 look even better. I had assured GP that the jack would be returned as soon as we were done with it. Both *my buddies* waited until the weather was good and changed the tires out. The crowds on the flight deck were really taking positive notice of her and the rest of the girls. Of course, #122 was getting monthly clean ups and the guys were forever spot washing the seagull dump off of her almost daily. I do remember one summer afternoon. It had to be up in the 80's and it was only 10:00 a.m. or thereabouts. I hauled the stuff up on deck to start cleaning her up. Since there was no other way to do this other than climb on her back and get on you hands and knees, that's just what I did.

I climbed up the maintenance handler, which by the way was a fight that I won. No more would the maintenance handler be pushed up the side of the aircraft. You could only see her from below and you could no longer walk underneath either, #122 had been roped off, much to the USAF and my great relief. I started to hose #122 down and as I was starting to soap her, I noticed that the deck was already getting busy. All I can say is if you have ever washed an airplane that's 102 ft. long, painted black on a hot day on your hands and knees, you know the meaning of HOT. When I would kneel down to start scrubbing, the heat from her back would burn my knees. No kidding, even if I wet my jeans down, that heat was crisping. The crowds were starting to build on the deck. Since I was busy trying to wash #122 and not fall off, I hadn't really noticed the crowds building around us. I looked down to see something like fifty or sixty people staring up at us. I was moving over to the mid fuselage, which required me working from the maintenance stand. Naturally, I was wet, and of course there were a few comments that I heard had something to do with Playboy magazine. Other people were appreciative and yelled up to me that #122 was looking great! People were really interested and that was a good thing. It also helped me keep going with the five hour job in the burning sun. What I didn't know was that *My boss* and *Museum Director* were watching the scene from mid deck., and just smiling away. I heard about it later from J, one of the former *Evil I* crew members who stood watch up in the island control room.

I finally got done somewhere around 2:30pm and was cleaning up to go downstairs. As I was hauling the hoses and buckets, I looked back and God! She looked awesome. What I liked even more was the crowds standing around her. People were just admiring her. It really made me and my back feel so good to see that. It wasn't until I got down to the shop, which was always chilly due to our rather weird air conditioning system that I realized how sunburned I was. In the coolness of the shop, it really started to sting. By the time I cleaned up and changed, I was cooking. I was so happy that I had gotten an air conditioned bus home even if I had the chills all the way. When I looked back from the bus before leaving 12th Avenue, #122 was standing black and beautiful on the flight deck. You could still see the crowds standing around her. I went home shivering and proud.

EA6B airlifted on the deck by a crane (Author collection)

An early photo of what the A-12 looked like prior to care and spy program (Author collection)

This was the first EA-6 that came off the production line at Grumman This was prior to the changes made by all of us (Author Collection)

The Grumman F-14D being lifted onto the flight deck (Author Collection)

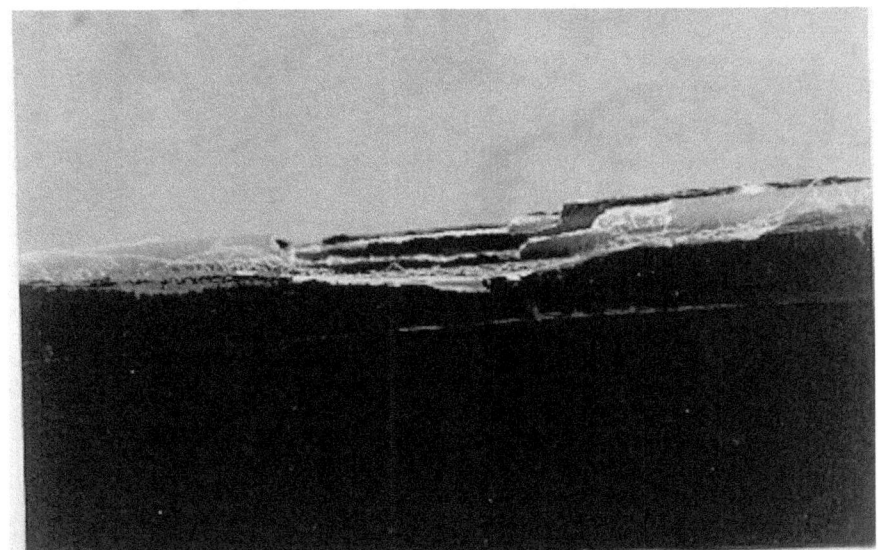

Some of the severe damage the A-12 suffered before our restoration program. (Author Collection)

The author in the "maintenance shop" (Author Collection)

JV comes to visit us at the Evil I with some friends (Author Collection)

The reverted tank sealant that took some 20 hours to scrape off (Author Collection)

All of the team for the F-14 move are lined up on the pier before the F-14 headed off on the tug down to the ship. I am the only woman on the team and I am on the left (Author Collection)

Fixing some tires on the aircraft when we finally got our maintenance shop going (Author collection)

Finally, CIA Article #122 in flight and heading for a refueling over Area 51
(Lockheed)

I am reassuring the A-4 that we will get her out of
the steam catapult shaft cover that collapsed under her.
(Author Collection)

The Grumman F-11-1 Tiger on the pedestal on the lfight deck. This is where we were moving the A-4 Skyhawk to when she fell into the steam catapult. (Author Collection)

The Vought F-8 Crusader on the flight deck after being brought in over the Hudson River via helicopter. (Author Collection)

The huge Lockheed P-2V Neptune. She was an amazing bird.
(Author Collection)

Up on top of the A-12 with my red hair shining in the sun, I am getting ready to clean the A-12 (Author Collection)

The next weekend, I heard from *My boss* and Museum Director. They were so impressed with how #122 looked, they felt the rest of the aircraft could use some of the same treatment. So, there I stood. I was trying to do the best for #122 and was doing just that, but the other girls needed help, too. After **Freedom Flight**, I couldn't say no . So, every weekend I ended up doing just about two airplanes or helicopters during the summer into the fall. And, those that didn't get washed like the girls down on the flight deck, (those were our WWII warbirds) got cleaned and waxed. I had a very strange incident happen down on the hangar deck. Our TBM Avenger, named for George Bush's aircraft in WWII, was up for a dusting and waxing. The day was fairly lousy out, it was raining, so I couldn't do too much on the flight deck.

 I decided to take on the TBM. As you know the TBM had a rather large nose and fuselage, in short, there's not much to hold onto, so *my first buddy* had warned me to be careful, because you really could get hurt with nothing to grab onto if you were falling. The step ladder I had was a smaller variety, not the usual painting ladder.

I had just gotten through dusting her and was starting to apply wax , Usually in a situation like this, I leaned into the aircraft for support and worked the higher areas this way. Something very odd happened. I know that FD, one of *Evil If*'s former crew members was sitting at the information desk just in front of the exhibit. He was talking to some visitors. I even had some visitors come over to see us, so people were around. I was waxing an area, when I lost my balance and **started** to slip, again with nothing to hold onto, I was going down onto that concrete deck. Then something weird happened, I felt someone or something grab me and force me against the aircraft's fuselage until I got my balance back. I was shaky when I stepped down off the ladder and looked around. I asked **HD** was there anyone around the TBM? He said he didn't see anyone and no visitors had gone that way in some time. I finished up on the airplane and met *My buddy#2* back down in the shop. I told him what had happened. Just then, *My first buddy* walked in and heard what I said. He told me to ask one of the other *Evil I* crew members, HD, about what happened. This I went right out and did. I found **HD** on the third deck and told him what happened. **HD** then told me about *Evil If*'s ghosts. It wasn't' the first time he'd heard of something like this. **JR** also one of the *Evil I* employees, had come to the shop later and told me some more about *Evil If*'s ghosts. While I was sensitive to objects and could read them, I never encountered a real, live ghost. *My first buddy* also had told me that there were many times that he felt someone in the shop that was looking over his shoulder, when he knew that no one was around. I was very thankful to the "gentleman " that had caught me. Next time I was on the hangar deck, I went over to the TBM and told him so.

Work continued on the ship and #122, but my back was becoming murderous. I knew that the type of work I was doing was exactly what degenerative disc disease hated. After a check up by my orthopedist, I was advised to curtail activities. How? How could I do that? It wasn't going to be possible, not at this stage in the game.

Here's a nice little story. I was up on the flight deck early one Saturday morning during the summer. One of the Navy ships had pulled into berth next to *Evil I*. Just a little boat, the aircraft carrier USS John F Kennedy was in port to visit for a few days of a Memorial Day weekend. It was 8:30a.m. and I was hauling hoses, buckets and rags up from the shop. The shop entrance was just off one of the catwalks on the side of the ship and the flight deck. The museum wasn't going to open until10:00 and so I thought I could get the F-14D and the EA-6B washed and polished before it got too hot. I got a short maintenance handler and pulled it to the side of the F-14. I could see the Kennedy's crews going out on shore leave from the 17 story high flight deck. The F-14D was just mid ship on the flight deck.

Her aft section was hanging over the fence and God, it was a long way down to the pier. As I climbed up and onto her back to start washing her, I saw about 12 or 13 guys in white uniforms coming up the main staircase. No big deal, the sailors always had access to the museum even when we were closed.

They showed themselves around the museum with our blessing. I sat down on the handler and pulled off my sneakers ad rolled up my jeans. I *started* to hose down the F-14, filled up the bucket, which I had on her right wing with cleaning stuff and started to clean. I was about 15 minutes into this and God knows I wasn't paying any attention to what was going on behind me on the flight deck because I was too worried about the 17 story drop if I screwed up and slip off the F 14's rather downward tilted back. It really was rather treacherous. I was stepping back and looking for the maintenance handler when all of a sudden I hear this applause. I looked back at the flight deck and saw the twelve Navy gentlemen in white standing there.

Apparently they had been watching the entire scrub down. I usually would whistle or sing while I was doing the scrub down, so they probably had a good time listening to me do Janis Joplin's "piece of my heart ". It turned out that one of the guys was the crew chief of the F-14's on board the JFK. He was smiling and laughing. He called out " You can wash any of my airplanes anytime! What a great job. I'm glad my guys could be here to see this, they can take a lesson!" I got some more applause and whistles as I came down the handler to see *Museum Director* who introduced himself and *My boss* and the crew chief and *started* to tell him what was going on. I could see *My boss* just grinning at me. I wanted to crawl into something. I knew that they would never let me live this down.

Well, *Museum Director* called me over and introduced me. I finally managed to get away and wanted to go back to finishing the airplane. Actually crawling into #122's nacelles was looking real good about now. I got a great salute from the troops before they left with My boss and *Museum Director* giving them a guided tour of *Evil I*. While I was embarrassed out of my mind, it still made me feel good.

While the summer was always busy on the ship, we had to get all of the heavy stuff done before winter set in. There were still lots of things going on in the winter. We checked the other girls on the deck all the time for ice and snow coverage, checked tires, tied down helicopter blades due t o horrible wind conditions on deck, they were always snapping. One Saturday morning, I was up checking on #122 to make sure she was doing okay. I was walking underneath her, when I noticed that the bright green almost neon crud was around the rivets again. Known as Kadena Kurd , it was a type of corrosion that was endemic to Blackbirds .

I went downstairs and got the Vaseline out , all of my "special tools" and went up on deck. I spent about 4 hours up there cleaning the stuff out.
The temperature that day was something like 33 degrees on the ground, so with the wind on the flight deck, it was a lot colder. When I finally couldn't stand anymore, I went downstairs to clean up. I couldn't feel my fingers, but #122 had beaten the Kadena Krud at least for a while. Anyone who's been on *Evil I* knows that seagulls are a mainstay. *My buddy#2* and I were on the flight deck. It was New Year's day. Yes, the museum was open, they had to clean up from t he New years Eve Ball the night before, so all hands were called on.

It had snowed a couple of days before, just enough to put a good 2" coating on everything. So, we were up on the flight deck checking things out and picking up left over wine glasses when we see a whole flock of seagulls land on the deck. That was nothing new in itself. They were always visiting us. Like at eight in the morning when they would drop rat bones and whatever else kind of garbage they could find on *My first buddy* when he was doing the trash. Anyway, this big, seagull, looked like a huge male. sits on top of #122 and takes the dump from hell. It just dripped all down the chines onto the ground. *My buddy#2* was ticked off as we stood there. " Look at the Freaking' bird. I just cleaned her off yesterday morning." The gull just sat there, looking at us, like we weren't more than trash. Then, he took another dump, staring at us while he did it., just to show us that he could.

I tried to chase him off by yelling but it did no good.. He just stared at us like we were an annoyance to him. My buddy was really ticked of by now. By the way this seagull had to be about twenty pounds and had mean beak. He really was big and arrogant. Next thing I see, *My buddy#2* is scraping some of the snow out from the back of the Corsair and making a snow ball.

Before I could stop him, he threw the snowball ad hit the bird square on. The bird just sat there, it did not move, at least not yet. *My buddy#2* was about to make another snow ball when the gull hops down onto the flight deck and started coming towards us. I grabbed *My buddy'#2's* jacket and pointed at this menacing, twenty pound, mad ass, aberration of a seagull walking towards us. I told *My buddy#2* to slowly head for the stairs and the shop. I saw the movie " The Birds" and I didn't want to upset Genghis gull any further. The bird made a lunge and grabbed *My buddy#2's* pant leg, . The gull was pulling it and trying to bite so at that, we ran to the stair case with this bird chasing us to the stairs. Before we left that day, we snuck back up to the deck just to peek out an see if Genghis gull was still there. Yep!! He was sitting right on the A-12 and dumping away happily with all his girl gulls around him, Guess we lost that argument.

It was a brutal winter. Every time we turned around there was either a snow storm or an ice storm. We could get virtually nothing done on deck. *My buddy#2* called me at work one day to tell me that the Grumman Albatross had about 3 feet of ice inside the cabin.

Apparently there must have been a hole in the wing or something that was letting the ice and snow in. There really wasn't anything much we could do, but we did try to melt it out with a welding torch, it froze right back up again.

Both *My buddy#2* and I had noticed *that my first buddy* was a little too quiet. Something was wrong, but he didn't want to talk about it. Finally one Saturday afternoon in March, he opened up. *My first buddy* had made the decision to leave *Evil I*. *My buddy#2* and I were stunned. It had to do with money and his wife. She was not happy with him working for less than she was making. The sad thing was he really loved what he was doing. But, the pressure was too much and he told *My boss* he was giving notice. *My buddy#2* had been working part time, it looked like he would be full time now. We hated to see him go. We were the "Nairobi Trio" and we worked so well together. So, *my buddy#2* and I said our goodbyes to our good friend. Even the airplanes were sad. I could tell.

My buddy#2 went on to full time, there was no help for him for quite a while. Money, once again, at the Museum was very tight. I was the only help that *My buddy#2* had for quite some time. There was tons of work, however. I could only come out on days off and weekends, but did the paper work from home and my office.

Ops was playing hell with everything. I took a week's vacation and worked out at the ship, washing airplanes, polishing, doing maintenance and paper work. My work load at the ship was becoming heavier. I also found myself shelling out money for tools and other things we needed because there were no funds for it. Every time we went to *My boss* or *Museum Director* for anything we got "We can't afford it". It really started getting to me. It seemed that Ops got everything they needed. *Ace* had five computers in his office, all of them with a different ISP so he could play on the internet all day, and don't think that he didn't. He claimed it was for materials they needed, so he could get the best prices. There was money for a new carpet in the hangar deck so the parties would look good. As I looked around, I started to see something was crooked. The Museum foundation members were all getting paid nice, healthy salaries and so was *Ace*. Museum Director, how ever, at one point in time had put his house up as collateral so the museum could stay open. That was the end of his first marriage. I found out things from other *Evil I* insiders. People had expense accounts and others were using the 501C tax free accounts for their personal use. It was becoming clear to me that there were abuses going on. Yet, there was nothing that I could do about it.

Chapter VII: Buddies, Airplanes Presidents and Nightmares.

Even with *My buddy #2* working fulltime and overtime, the work load was tremendous. One of the first nightmares we knew we were going to face was going to be the dreaded "Fleet Week". This was an annual affair run by the *Evil I* and the City of New York, or visa versa. I forget which. The planning would started usually around February, when we would start finding out which ships would be coming in for the festivities. Depending on how important the Memorial Day weekend was going to be usually dictated how horrendous our jobs were going to be. It was going to be the mother of horror! My boss called a meeting to give us the schedules and breakdown of events. The big news was that President Clinton, Bubba himself, was going to be on our flight deck. I wasn't up for Clinton, but I'd make sure that Marine One would have a good time! Screw the President, and I don't mean that literally. Chaos *started* to reign supreme early on in the show. By March, the secret service was already screening everyone involved with the event. Since I was on the flight deck, the call went out to go and get interviewed. I have to confess, I was worried about this.

I knew that *Museum Director* and *My boss* weren't going to be there during the interview, but I was afraid that something might get back to them. I knew one of the questions was did you have any connection with the military. I was trying to figure out how to do this. Do I lie to the Secret Service. I didn't have to wait long to find out. I saw the guy who was going to interview me. He seemed like he was running the show. The secret service was also beginning to make us insane with their myriad requests. *My boss* told me that I was going to be next after *My buddy#2*.

Great! The guy looked like the guy who played the Assistant Director of the FBI on the X files. Bald, stern, rod up his butt, you know the type. I felt lunch coming up when I got called at 2:30PM, that afternoon. They were using *Museum Director*'s office to do the interviews. My boss walked me in and introduced me to this guy. You know, I never did get his name except for Agent M, but I do remember the little ear piece he had on.

Anyway, the Secret Service asked me to sit down. We had already filled out forms, which he had in front of him. First question, ever been arrested? No. Second question, how long are you at your present residence? 15 years.

US Citizen? Yes. Were you in the Military or did you work for the Military. DUH?! Well, I hedged, yes. You do or you don't which is it? Yeah well, I do, but *Evil I* can't know about it. He asked for my id's and I gave him everything I had including my Wright-Patterson AFB Museum ID. Why can't *Evil I* know? I attempted to explain, but he picked up the phone and called Wright - Pat Base security. I sat and listened, while he checked and confirmed that I was attached to the National Museum of the USAF (TM) Well, at least that went down okay. He asked me again, why can't *Evil I* know? I tried to explain it and the nature of what I was doing. I told him that if *Evil I* found out, it could jeopardize relations between the National Museum of the USAF and the *Evil I* and the Pentagon won't be happy if *Evil I* gets ticked off. He then made me explain everything I did, what I had access to, what I knew about the flight deck, airplanes etc. It went on for just about forty five minutes. I was sure I wasn't going to get the clearance when I finally got of there. To top it off, this guy was really ticking me off. So, I went back upstairs after the interview to do some work and burn off the stress. I was getting ready to get my stuff and head home after putting in a few hours on the flight deck when *My boss* came into the shop. He told me that the Secret Service approved me for access to the flight deck and to work up there during the presidential visit

I was one of only a handful of civilians that would be allowed up there when Bubba arrived. I was surprised but kind of glad too. I really would have hated not getting the chance to see Marine One up close. I left the museum with *my buddy#2* that afternoon and we talked about the whole thing on the way home. The tough part was going to be the airplane moves. That's what we thought then.

Time was getting tight for Bubba's visit and we were scrambling for stuff. My buddy was searching for tow bars and I was looking for Plexiglas for the Shawnee. I had taken another week off from work to accommodate this madness. There was no other way. I was getting out to the museum, about 6 am and hanging out till all bloody hours. The Secret Service had now moved in and were "arranging" things. The push was on .

My buddy#2 and *My boss* were trying to figure out how to accommodate something like 3000 people on the flight deck at one time AND have the appropriate space to land Marine One in comfort. OPS had built a set of bleachers for the mid deck. It sort of solved the problem .

I was now told that only military personnel would be allowed on the deck to greet president Bubba. Joy. As you can see my response was not overwhelming happy. So between trying to arrange helicopters and aircraft, we now had to put up with Secret Service telling us what to do. A couple of fights had already broken out below decks because OPS was not permitted to be on deck when Marine One arrived. Usually we have *Ace* on deck pretending that he knew what to do to wave in a helicopter. This time the *USS America* would be doing the honors. You can bet *Ace* was pissed. So I heard he had a real dust with Old Baldy as I called him.

We had a problem. So, what's new? Early one morning we realized that we couldn't move the Huey 1-M and the Sea Stallion, we had no wheels. They hadn't been moved since they were brought on board.. So, *My buddy#2* and I went wracking our brains to figure out where to get helicopter wheels. I remembered that there was ANG base out on Long Island, and they had Hueys.

I think I called SB of the Air Force Association. He was the former commander and had just been out to *Evil I* to seek our help in getting *Evil I* to host AFA meetings. I was a member of AFA and had been doing the newsletter for them, I was sure SB would help. I called him and 20 minutes later, he called back to tell us to come out to Republic and we could borrow a set of wheels. *My buddy#2* and I were pleased to say the least.

The next morning at 7:30, we were off on the Long Island Expressway to Republic Field to get the wheels. My buddy#2, just so you know, is a musician, he plays bass. You already know that I sing. When you add the traffic, fatigue factor for both of us and the LIE at 10 am, rain and just boredom, you get the rather loud and rockin' Aircraft maintenance/restoration band. *My buddy#2* revved up the car stereo and turned it to the hard rock station, the rest is LIE history. Every song that played, we sang loudly. I think we actually go the traffic moving because people wanted to get away from us. Personally, I think we sounded great, besides we had our windows rollup, not that it made much of a difference. At least it helped the time pass. By the time we finally reached the airport and the ANG, we were pretty well sung out, but happy. Even the pouring rain didn't bother us.

We walked into the ANG hangar and were just astonished at al the beautiful helicopters they had, all Hueys' all 1-M's. The guys were working on them, either electronically or mechanically. We were both in heaven watching all this. We picked up the wheels, just about got the tucked into the trunk of the car and were back on the road to *Evil I*. It took us almost two hours to get back in that rain. When we finally got to the ship, my buddy and I were already exhausted but no rest for the weary.

We unloaded the wheels and were met by My boss down on the pier. We had to move the helicopters right away because the Secret Service was coming in to inspect what were doing. So in the pouring rain, My buddy My boss and I trudged up to the flight deck with the wheels to move the helicopters. To say the least, it was suicidal attempting this in thunder and lighting but that didn't stop us....noooooo. We hooked the wheels up to the Huey and *started* to move her further up the flight line, followed by the Shawnee and the Sea Stallion. By the time we got downstairs, all of us were soaking wet. I didn't care what My boss wanted, I wasn't doing another thing until I got a hot cup of tea. Screw the Secret Service. The minute I said that Old Baldy stepped into the shop. Know what? I really didn't give a shit. I'd just about had enough of the Secret Service and their "inspectors". My boss and My buddy looked a bit nervous. I stared Old Baldy down while I was dripping all over the floor. I excused myself and went to the hangar deck cafeteria to get that hot cup of tea.

When I came back or when I was on my way back, *My buddy#2* was also on his was to the cafeteria. I asked him what was up with Baldy. *My buddy#2* said he thought *My boss* was going to die after I left the shop. *My boss* was almost apologizing but Baldy told him not to worry. At that point, *My buddy#2* said he left to come down here , just to get out of there for a while and warm up.

My buddy#2 was showing some of the same concerns I was, as we listened to the thunder outside. These guys were really ticking us off. The past week was hell and this week leading up to the Presidential visit was getting a LOT worse.

My boss finally came down to get us about 20 minutes later. Asking if I had cooled off, I told him , no just warmed up. He just laughed and then proceeded to tell us that we had to move the A-4 today, after the rain stopped. *My buddy#2* and I just looked at each other. Where? The answer was over by the F-11 by the staircase. The Secret Service felt that the area was too open. They wanted it to block the view from 12th Ave. Okay. *My buddy#2* and I passed Baldy on our way back to the shop. He just sneered at us as we walked by. I was already cursing under my breath. That guy just rubbed me the wrong way.

The rain finally stopped and the sun was attempting to poke out. It was already 4:30 in the afternoon. The deck was wet and slippery. Some of the other *Evil I* maintenance people were helping us bring the chains and tow bar and the rest of our gear up while *My boss* revved up the tug. The battery needed charging so it was going to take a while, at least an hour before we could get going. The Museum would be closing so at least we would be able to do this move in peace, so I thought.

My buddy#2, and **ND** who was our new part timer in the aircraft department . **ND** worked as an AP for United Airlines. He started prepping our beautiful A-4 for the move to the other side of the flight deck. She was now just forward and to the right of #122. We had plenty of room for the move and wouldn't interfere with #122 at all.

My boss finally got the tug up to sped and was over by the A-4. I have to tell you she was one of my favorite on the ship. This little A-4 Sky Hawk was a Vietnam Vet. The story went that her p[pilot had put her down on a field in Vietnam which had belonged to the French. She had some mechanical problems. Instead of bailing out over the jungle, he felt he had a chance landing on the airfield.

This he did, only to be captured by the Viet Cong. The Viet Cong took possession of the aircraft. As the story goes, I don't know how true this is. There was a raid to get the pilot out and the aircraft was recaptured by the U.S. Army. She definitely had personality and I really loved this little girl.

The museum was closing for the day, so as the last stragglers left the flight deck, the guys *started* hitching the aircraft up for the move. I was over by #122 watching #122's wing for clearance, while My boss was staring the tow.

As he pulled out, he got up a prey good head of steam and *started* moving the A-4 out, to #122's right. Just as he got over the catapult plate, which was made of thick glass (this was done so that visitors could see the working cables of the catapult) a kid comes up the starboard cat walk and walked out in front of My boss.

 I was watching the A-4 and #122 so My boss was already forward of me. He stopped short, not to hit the kid and the A-4, all 25,000 lbs of her was sitting on top of the glass. There was one second and then all we heard was "CRACK! The A-4's right wheel and strut went through the glass and fell into the catapult well. Her wheel was actually sitting on top of one of the cables.. My boss *started* yelling at the kid and the security guard (I use the term loosely) who came up after the kid. **ND**, *My buddy#2*, **MT** and I ran to the A-4. We were screwed and we knew it.

Guess who waltzes up the scene? You guessed it, Old Baldy. *Museum Director* came on deck because of the commotion, which had guys running to the shop to get the jacks. As *Museum Director* approached the chaos, all I hear is "What happened Jeannette!" At which point I was freaking out because Old Baldy asked would we be able to fix this soon? That was the first time I almost hit him.

Museum Director pulled me back and told everyone to calm down, we'd get it fixed. Okay, so we huddled by our poor girl, trying to figure out low long the cable was going to hold her before it snapped. If the cable snapped, then we were doubly screwed because the strut would snap, then we'd have one helluva time extricating her out of there.

None of the jacks that we had were going to be enough. We were very fortunate that *My boss* had a friend in the Police Department . He'd helped *Evil I* out once before. He worked with the Subway team and they possessed a twenty ton jack which was used to more derailed subway cars. They really looked like pillows until you put the juice on them. If they could lift a subway car, they could lift the A-4. Okay, the call was made.
In the meantime, we worked on securing her to best we could. It was late April so at least we had a couple of more hours of daylight. I felt so bad for her. At least it kept me off Old Baldy's neck. The cops showed up with their magic equipment. The problem was they weren't used to lifting airplanes. The typical scene occurred. Everyone was talking at once. They were making me nuts, so I stepped back and just watched. Then I saw it. The toughest part of any Navy aircraft is the undercarriage, right by the wing box. It had to be built that way, because trapping on a carrier wasn't exactly the most gentle way to land. Let's put it this way, if an F-16 attempted to trap on a carrier, it would soon be sans wheel carriage and belly. It would rip her guts out. So, I went over, tapped *My buddy#2* on the shoulder.

He was so mad and trying to control it. Old Baldy was standing by watching. I said to him instead of lifting her and setting the jack up mid wing, set it up under the wing box, right by the struts.

It's the strongest section and we won't have to worry about damage and balance. **ND** looked at me and *started* to shift the pile of wood, the cops there followed suit. The guys were loading and I was running around looking at the balance. I finally stepped over to her left side, hand on her fuselage and waited for the lifting to start. *My buddy#2* and the cops got the two steel plates from below deck to slip under her when she came out of the hole.

Slowly, slowly, she started to come up. I was praying the struts would hold, rubbing her belly a little more. Finally, she was up enough for us to slip the plates underneath. My buddy and the cops got it under her while **ND** secured. I pulled the chock from behind and out little A-4 was safe. Bet that made Old Baldy pretty mad, we weren't going to screw up his precious schedule after all.

Everyone finally *started* breathing again and we pulled the A-4 over to a safe spot. The sun was on its way down and the deck was starting to slip into darkness. We still had to move her into place.

Don't think in your wildest dreams that there in anyway this was going to be easy. We still had to maneuver her to just about mid deck, move the 2 helicopters over and then park her behind the F-11 pedestal. By the time we got her to mid deck, all of us needed a break.

The guys went downstairs to get some beer and I went to get a soda. *My buddy#2* and I stayed up on deck while we took this twenty minute break, talking about what happened. *Museum Director* had come up on deck to check things out. He asked was everything was okay?

Yeah so far, I said. He was trying to be calm about the whole thing because he knew the rest of us were stressed. He told us though that he was proud of how we handled the problem. *My boss* had come back topside with the other guys, Heineken in hand. We were ready to slip the A-4 into her slot. The deck was getting darker. I never realized just how dark it could get up there. All the lights we had were the ones strung from the island. So, not really much light at all. It was already fairly dark and the shadows were the ultimate bitch. You really couldn't gauge any distance accurately and we only had inches to play with. The trick would be to park the A-4 behind the pedestal with the F-11 so that you couldn't see the President's back .

When you looked out off the deck, you suddenly realized just how wide open the podium would be and we had lots of roof tops that we were looking into. Much as I hated to admit it, Baldy had a point security wise.

So, now it's just about 10pm. We're trying to get the A-4 snug behind the F-11. We had only 3-4 inches to play with before she went over the edge of the deck, 17 stories down. We did have a fence tacked up there but that didn't mean much when you had all the weight behind you that she did, it would be easy to have something happen. After moving the A 4 back and forth inching in a little, pulling out and moving another way, we finally got her to a place that looked just like what the Secret Service wanted, or at least that's what we thought.

WRONG! It was really getting super dark on deck and all we had were flashlights. So, everyone was taking a breather for a few minutes where here comes Baldy up the stairs. I just felt it coming. I think *My buddy#2* and *Museum Director* did too. Baldy comes up with his super, duper flashlight, takes a look at where we had the A-4 walks back on the flight deck towards the stern, stands there and then walks back to us.

I was standing between *Museum Director* and *My buddy*, both rather large gentlemen. Baldy said to *My boss* that he didn't like the position. *My boss* looked really stunned, he asked him WHERE he wanted us to put her?!? Baldy pointed to a little hole that was between the Huey and the A-4. Close that up, he said. *My boss* was ready to blow. He was usually a very quiet guy, but when he was going to let go, I could tell. *Museum Director* asked Baldy could it wait until the morning, when we had more light. Baldy says," No Need to have it done tonight. No one leaves until its done right." I was ticked off now. "Why don't you take off your jacket and give us a hand." which flew out of my mouth with venom attached. To which Baldy replied. " I don't get dirty, that's your job" With that, I launched at him like a stinger missile. *Museum Director* caught me and *My buddy#2* grabbed from behind. I was inches from his face. *My buddy#2* pulled and threw me back and *Museum Director* got in front of me. My buddy slammed me up against the pedestal under the F-11. *Museum Director* actually told Baldy off! I was amazed! I knew I would catch it later for snapping at the secret service. However, *Museum Director* let him know that he was asking us to do something very dangerous on a very dark flight deck. Baldy told him flatly. It gets done tonight or no Mr. President.

My buddy #2 still had me pinned to the pedestal. I watched mister Secret Service pass to go back downstairs. I really wanted to rip him up. *Museum Director* walked over to me. Here it comes I thought. All I got was take it easy and lets get this done. My b finally let me go and *My boss* made me promise not to go after Baldy. The guys were laughing but I think they wished I would have slugged him. I'd be in jail for that, but it would have been worth it. Here it was almost 12AM, we'd been working from 6 AM and we still had to jiggle the A-4 into position.

By 1am, Baldy pronounced it good. *My buddy #2* and I headed down to the car to go home. We had to be back by &AM. Bubba was due on the flight like the next day. We showed up at 7 the next morning, or should I say that morning. When we got to the deck, I was truly amazed that we hadn't dropped the A-4 over the side on her head. She was just barely on the flight deck. Even when we put the fence back, she was bulging out on the side. We were running crazy. We had to move the F-14. *My buddy#2* and I went down to get the tow bar, which was big and heavy. We also needed to tighten the screws and do some other stuff to it. Don got called so I stayed on the hangar deck to service the tow bar.

It was madness! *My buddy#2* and I were getting pulled in all directions. Baldy and his buddies were everywhere. I had to go up to the shop to get a different wrench when My boss stopped me. He told me that from here on in, only *My buddy#2*, **ND** and I would be allowed into the shop. Don't send anyone down for tools, because the Secret Service will not allow them in. Great.

 I had a bunch of heavy stuff I needed to take with me along with the wrench. My buddy called down and asked for some screw drivers, cables for the tug (which was on the flight deck and having issues of its own), the drill and various other assorted things. So I took what I could, dumped it on the flight deck, got my wrench and went back to work on the tow bar. When I got there I see 4 guys, paying customers, standing by my tow bar admiring it. When I jumped over the bar and went to work, I got some very odd looks. I kept working and the guys kept staring. *My buddy#2* finally came down to get the bar and take it up in the bomb elevator. He was laughing at the crowd of admirers I had attracted. We both took an end and walked down to the bomb bay elevator. This elevator is an actual WWII, never fully restored, sometimes functional lift to the flight deck. The Ordnance for the aircraft was brought up to the deck this way. It was a temperamental thing at best

We rode up and deposited the tow bar on the deck. The tug was running, first we had to move the F-14D, then the EA-6B, after that we had to move the E-2B Tracker so Marine One could have some more room. We were pretty much exhausted when we finished but everyone was where they were supposed to be. I swear, as I was walking back from the bow, I looked at #122. She looked like she had this grin on her face and was laughing at me. We had to take the tow bar back downstairs, so **ND** and *My buddy #2* and I hauled it over to the bomb bay elevator.

There was an Ops guy by the name of **D**. It was his sole job to operate this freaking elevator. No one else could touch it. **D** was a true Ops cowboy. He loved to bounce people around on that thing. You've got to remember it was nothing more than an open platform with cables running through each corner. It was dangerous because it rode right through the deck. If the cable slipped, you flew your way down to the engine room. There was nothing to hold on to. We loaded the tow bar on and we wanted off on deck 3 so we could stow it there. **D** was cute with the first deck. As a safety measure, you had to stop at each deck, then move to the next. Well, we all bounced and almost lost footing.

I had my foot under the bar so I was able to grab that. The elevator *started* to move again and then as we all looked at each other, it *started* to speed. You know that sick feeling you get when a elevator goes into free fall? It happened to me once in the Kodak building. The freaking thing dropped from the 47th floor to the 23rd floor before the brakes kicked in. I was sick for weeks. I also walked down from the 23rd floor and refused to go back into the building. Anyway, this thing *started* flying. We grabbed each other and it finally bounced to a stop on deck 4.

One of the cables slacked. Since I was standing near the corner of that cable, it slacked a fell over me. You never saw 2 guys move so fast. My buddy and **ND** grabbed me out of there and threw me down before the cable could tighten again. When the platform stopped, **D** was yelling and laughing down at us. I went nuts, jumped off onto the 4th deck, screaming back at him and calling him every curse I could think of.

My buddy#2 and **ND** came after me, as I ran for the stairs. I was going to kill this bastard, but not before **ND** and my buddy shoved a few tools up his butt. By the time we got up to the flight deck, *My boss* was screaming at **D** too. It was 4 PM. Would this nightmare ever end? When we finally calmed down, I was totally exhausted, disgusted and screw it, I was going home. My buddy#2 persuaded me to wait and we'd all leave in another hour. I helped stow the tools , got washed up and waited down in the shop.

The place was crawling with Secret Service below decks. I figured I'd do some paperwork while I waited for the guys, when Baldy shows up.

I wanted to ignore him, I was in no mood, but he was actually almost pleasant for a change. He asked me if all tools were accounted for, made sure no one else had been in the shop, and actually to told me to take it easy. He saw how hard we were working and by Tomorrow at $, it would all be over. Did I have to wait that LONG!?!?! Well, bolstered by Baldy's concern, I was really sick now. Finally the guys were ready to leave. My boss handed us our passes for tomorrow. Without them, we didn't get on the ship . He reminded us, no jeans, shirts and ties for the guys, slacks and skirts for the ladies. (Was I a lady? I'd forgotten) and yes, be here by 7 am. I looked at the passes, was thanking God that after 10:30am, when Bubba hit the flight deck, my job was over. Yeah right, and I had a bridge to sell you.

The morning of the big day finally arrived. 7am and *My buddy#2* and I, with Dunkin Doughnuts and coffee in hand, pulled up to the *Evil I* gate and got our assigned parking pass over at Circle line. As we went through the metal detectors and showed our id's we met *My boss* waiting on the other side.

The place was screaming with Secret Service. I couldn't believe how much Brass what there waiting in line to get in! White Naval uniforms, Blue USAF uniforms, Green Army and Marine uniforms along with other VIPs. Incredible!

Below decks, which were now off limits to anyone, along with the hangar deck, Secret Service were scurrying around. The troops were being led up to the flight deck and getting into position. There were lots and lots of military, some from Canada, Britain, Israel, Spain. *My boss* got us up to the flight deck . It was there *My boss* told us to start searching all the aircraft, a Secret Service person would be with us as we did it. Anything they wanted, they got, just give it to them. What's new?

On my way to #122, I caught a glimpse of *Museum Director*. He was a nervous wreck. I almost wanted to laugh but I had to start checking #122 , the Corsair, the Shawnee, the F-14 and the A-6. My buddy#2 , and **ND** were getting the rest. The deck had to be cleared by 9AM. It was then, after we finished checking the aircraft that *My boss* told us we were getting a special job. Great, now what? The Secret Service came over and showed us this curtain that had been rigged by the podium, right in front of the F-11 pedestal. When the President got near the podium, My buddy, **ND** ,**MT** and I would roll this curtain down and tie it off on the bottom of the fence. For this, I was not prepared. The president would not speak until this was done. Great. Okay, with everything else we had thrown at us what was a freaking curtain?

We all tried it out once and the SS seemed satisfied. My buddy, **ND**, **MT** and I just sort of looked at each other. No one,. mentioned this to us. No One also mentioned that we were assigned VIPS. My buddy pulled 2 Navy Public Relation guys, **ND** got two Army, **MP** got a British someone or other and I got two USAF Public Relation guys. Go figure. Well, the big moment was arriving. The troops were being led up on the deck. The Navy of course was front and center. The PR guys also made sure that all the lady Navy personnel were up front to greet President Bubba. The weird thing was it looked like they were doing it on purpose. Well. Bubba loved the women, so I guess that had something to do with it.

9:50AM approached and I could see the Navy guys on the rear of the flight deck on the radio talking to Marine One. We stowed our PR guys in there and went to wait by the podium to drop the curtain. While I was standing there, I noticed 6 guys in black, turtlenecks, black pants and black gloves with bullet proof vests, come out with these long thin cases. You didn't need an imagination to figure out who these guys were snipers. That really rocked me. I suddenly realized ALL of us were in the line of fire if anything went down. I think we all get the message at the same time because

My buddy, **ND** and **MT** looked nervously at the guys who were set up on the stair case directly below us. *Museum Director* came over to us basically to say " don't fuck up." He was so nervous , it was almost comical. He had the gift jacket with the *Evil I* patches on it to give to Clinton along with some plaque or something. All I wanted to do was get out of there. I was hoping no idiot would tell my boss something. I have to say the SS really had the place locked up tight and yes, we were the only civilians up there.

Marine One was in view and hovering over the deck as the Navy guys from America were bringing her in. She really was awesome. I could feel my girls looking at her in envy. She was finally hard down and the deck waited in breathless anticipation for President Bubba to arrive. Out he stepped onto the deck and the crowds started cheering. *Museum Director* started sweating. We were all getting worried he'd blow it. Anyway, Bubba was swaggering up the deck and heading for the Podium. He had on this great suit, rally sharp but I've got to tell you, I really didn't like his swaggering attitude. He looked like some old rock star instead of the President of the United States.. He made it to the podium, took another couple of bows, while we dropped and tied off the curtain. We then walked off to our PR charges and watched the rest of the floor show. I was standing with my two USAF Majors.

I've got to tell you that while the soldiers were impressed and whooping it up, I really didn't see too many of the upper ranks going at it. The guys saw the same thing. It was weird. I felt my two charges were just doing what they had to.

Clinton and *Museum Director* did their little *Evil I* welcome then Bubba stepped up to speak. The girls from the Navy were carrying on and Clinton was loving it. He looked at every single one like they were the only one on the deck. I really felt like this was rock concert, not the auspicious occasion that it was supposed to be. He finally finished his little speech, then went out with the Secret Service behind him to meet the troops. I was standing under the Shawnee with the 2 USAF guys. Clinton came over and since I was in the first row in front, I got to look right into his eyes. Yup, he was eating it up. It was strange, but I had no impulse to shake hands with him. (I wonder why?) so I just nodded and smiled back. The USAF guys climbed over me and shook hands with him, while the Marines and the Army just stood by. I really felt awkward. Well, it went on for about twenty minutes and finally everyone *started* down for the hangar deck reception.

Only the foreign brass made any attempt to shake hands. It My boss herded us down there for pictures and stuff since real estate magnate (*Evil I's* owner) was down there. I just suddenly got the feeling I didn't want to be there, so I grabbed a coke and headed up to the flight deck to hang out with #122. It felt good to relax for a few minutes because Saturday, the next day, we'd have to start moving aircraft around the deck again. My buddy escaped from the hangar deck pay bossy and came up on deck. We asked the guards if we could look at Marine One. They were cool and let us get close enough to see out reflections on her brightly polished side. She really was awesome.

I think for me, she was the best part of the day. I got myself out of there before President Bubba left. I told *My buddy#2* to tell *My boss*, I was having an asthma attack and needed to get home. So I left without anyone else knowing and headed home. Saturday morning, I got to sleep late. I didn't get to *Evil I* until 8 AM. We had to start moving airplanes back to their original positions and clean up the flight deck. I don't know why, I just didn't feel much like talking so I just pushed and pulled and turned when told to. *Museum Director* and *My boss* were flying high when they got on deck that morning, before the Museum opened for business. We were already moving the A-4 back to where she needed to be. I was hurting madly and just wanted to get finished and go home.

Even *My buddy#2* couldn't get me to snap out of it. *Museum Director* and *My boss* came over to tell us how wonderful it all was, so I decided to hide behind the Corsair. That is until *My boss* found me. The questions started, where were you, *Museum Director* was pissed he wanted to show you off…etc etc… Then *Museum Director* came over, and it was the same stuff. I just told them that I wasn't feeling well and I felt an asthma attack coming on. I just needed to get home and be quiet. I don't think they really believed me, but it was going to have to do. It was the best I had. Besides, there was going to be a demonstration on deck at 1PM and we had to be finished moving so OPS could do their act.

There was going to be a Navy Seal Rescue demonstration landing on the aft flight deck and of course, *Ace* from OPS managed to take control of this one. Whatever. I just wanted to go home and take painkillers. About 12noon, we had just about moved everyone back into place, but we still had to move the E-2B aircraft back. With her big, pancake top, she was always ungainly to deal with. Things were taking time and I could see OPS laying out the stands and ropes for the visitors so they could see the show. Unfortunately. It was right next to where we were working on the E-2B Tracker. 1 PM rolled around and this big Navy Ch-46 Sea Knight was rolling in too. The crowds were large and very , very close to us. I was up at the E-2B's nose waiting for the next order to move. We really weren't paying too much attention to the show because were worried about the crowds next to us and safety not to mention the E-2B.

Quite suddenly, I felt the downblast of the Sea Knight rotors and looked up to see a Navy Seal lowering down on a rope to the deck. The Sea Knight was hovering right over us. You have to remember that *Ace* was calling her on the radio and the idiot put her right over our heads. Well, what happens when an airplane gets into a wind, she wants to fly and considering the E-2B wasn't chained down yet, that's just what she did. The guys were working on the wheels to loosen up a couple of frozen brakes and that is when Miss E-2B made the decision to take off. I felt the E-2B lurch and pull up off away from me. I grabbed onto the nose strut, which was rising and started screaming (over the helicopter noise) She's gonna GO!!! It was only a second, but it seemed like forever as I felt myself being pulled up by the E-2B. Finally I heard a Holy Shit! from behind me as the E-2B started to shift towards the crowd. Everyone jumped on whatever they could grab to hold her down. The E-2B was rocking. **ND** jumped on top of me as I held onto the nose strut, because I didn't have enough ballast to hold her down. *My boss* got 2 other ops guys to grab chains and he ran over to *Ace* to tell him to wave the helicopter off. *Ace* turned and looked. (understand he was the "safety" man during demonstration and he was busy taking videos of the demonstration). *My boss* started waving frantically at the helicopter, who wisely decided to move off down the deck. I felt the E-2B settle and **ND** slipped off my back. I don't know what happened to me, but something in me snapped. I was raging. *My buddy#2* saw it and followed me down to the shop. Once there, I completely busted the shop up. Everything went, garbage cans, books, tools, tires, all of it. I had had it. I remember *My buddy#2* leaning on the doorway watching me trash the place. I guess I knocked myself out because I just sat down finally, shaking. *My buddy #2* just looked at me and asked if I was okay. I told him straight; "Working here is like working on a suicide mission !!" I grabbed my stuff, went to the ladies room, cleaned up and went home. I refused a ride home from *My buddy#2* and took the bus. As I calmed down , I realized something else, I couldn't feel my left leg. It was numb and I was hurting massively. I guess if I had to look anywhere for the real bottoming out of my spine, that afternoon was the start of it. I didn't show up at the ship for a couple of weeks

Chapter VIII: What Helicopter?

It seemed like helicopters were becoming a big part of my *Evil I* life. If I wasn't trying to find out who owned the Hueys, it was finding parts for the Shawnee, or tying back helicopter blades on the Sea Stallion. There was one special hard case that had been tossed my way. I got a phone call from *My buddy#2* at work, asking me if I knew anything about a CH-57 helicopter that was the only surviving vet of the Iranian Hostage Crisis. He told me that he heard they were going to scrap it out. She was the last survivor of the mission. Well, I wasn't going for that. I made some phone calls, the first to the naval historian **MW.** I got the whole story and refused to just let it happen. So I badgered him and the Naval Curator and basically anyone who would listen that just because the mission was a failure, you couldn't just scrap the last survivor out of the world and forget all who died during the mission.

To make a long story short, I won the fight and today that restored CH-57 sits as a remembrance for the JFK navy Seals Training Center in Cherry Hills, North Carolina. Yea! One for me and ZIP for Revisionist History.

This brings me to another infamous helicopter. I say infamous because through no fault of her own, she was used badly. It all *started* out with a phone call at work one day. I got a call from *My buddy #1* (who was leaving the job that week),about 1:00 P.M in the afternoon. He called to tell me about the "new"" helicopter that showed up at the pier that day. *My first buddy* asked me if I knew anything about it. I said no, I didn't. Funny thing was *My first buddy* didn't know anything about it either, neither did My boss.

Well, I really didn't give it too much thought. I was kind of busy that day and basically asked *My first buddy* to let me know if he found out anything to let me know. I did ask him, however, what kind of helicopter it was. He told me it was a Huey 1-M, latest class, beautiful condition, rocket pods still on her and she came down from Buzzard Bay, Ma. I asked him if the helicopter was de-milled. The answer was "No." Very strange. No aircraft of helicopter ever went to any museum without being demilled in some shape or form.A week went by, not too much more was said about it and the helicopter sat by the pier next to *Evil I* on a barge. One morning *My first buddy* comes into work and the barge is gone.

My first buddy asked around the museum and no one knew anything about it. Finally one of the guys from the submarine Growler told him that they had towed her over to Staten Island early that morning. On whose authority *My first buddy* asked. Well, the guy didn't know or didn't' want to tell. In any event, the helicopter was gone as mysteriously as it had shown up.

Neither *buddy#1* or *My buddy#2* or *My boss* or even me for that matter, knew what the story was and no one was talking. The whole thing just seemed strange. A few days later, I happened to be talking to **PL** our GSA representative in San Francisco. He had been helping me locate some parts for the aircraft and had been helping me try to find out the origin of some of our exhibits. I was just about to hang up from a nice, pleasant conversation when he asked me" How's that new Huey 1-M we just sent you?" I froze like a deer in the headlights. I managed to get out " Oh, ...yeah.... She's just fine..." "When are you guys getting her up on deck and on exhibition?" I just kept going. " Oh real soon, we're just cleaning her up." Finally he hung up after being reassured that the Huey was okay. Yeah right, like we even knew where she was!

I called *My buddy#1* immediately . I was in a panic. I told him I had just lied to a government official and I wanted to know where that freaking helicopter was right now!! He managed to get me to calm down and said he'd go and talk to *My boss* right now. He would call me right back. About 15 minutes later, a stunned *My boss* called me back and asked what happened. I explained to him that I had told the GSA that the helicopter was alive and well at *Evil I* to save the museum any questioning and problems, but I had to know what was going on. We agreed to meet the next day, Saturday.

I went over to the ship the next day in the morning. We all stepped out and went to H&H bagels to get some coffee and bagels and figure this thing out. *My buddy#2* and I decided that we'd go over to Staten Island and see if we could find her. There weren't too many places you could hide a helicopter in Staten Island. She had to be somewhere. While waiting for news I called **TB** in the PC office on the quiet, just to see if they had any ideas. The main point was one—I had lied to the GSA. Even if it was to protect the museum, second, -- we could try to find her but eventually we'd have to confess and tell the GSA and yeah, it would go to the Inspector General of the GSA Anyway, I'd be damned if I'd go to jail for some dishonest bastard on the *Evil I*.

Chapter IX: Road trip to find a missing bird

Saturday rolled around. It was a nice, bright, cold November day. 6:00am to be precise, when *My buddy#2* and I rolled up to Dunkin doughnuts to get some supplies, mostly jelly doughnuts for the road trip to Staten Island. *My buddy#2* even brought a tape of helicopter noise to play to see if we could coax her out of hiding. Staten Island isn't that big of a place. It sports one armory, a historical fort known as Ft. Wadsworth and of course, the now defunct Homeland base, which was used by the Navy as a New York port. There was a fishing marina on the other side of the island to explore too. *My buddy#2* and I thought of every angle. Where would someone hide a helicopter? It didn't make any sense. We started at Ft.Wadsworth. The plan was I'd get out of the car, being a nice, cute redhead, not too bright and ask the guards if there were any helicopters around here? We'd heard that there was a new exhibit. The guard looked at me like I was nuts and said that there were no helicopters here and never were.
 Next, we'd drive up and down the Homeland docks. The place was huge! It was also freezing. Every couple of piers we'd stop and see if there were any barges or people around to ask. Lots of water, cold deep water but no barges and only two guards who also thought we were nuts, asked us for ID then told us to move on.

Our next stop was the armory. It was a VFW along with a National Guard unit. The place looked like a small castle, duly fenced in with appropriate warning signs that deadly force would be used if unauthorized entry were attempted. The Armory had a big fenced in yard around the back with the usual array of National Guard trucks and jeeps and garages, which could have housed a helicopter. Since we couldn't go in, we asked if we could be allowed to look around the place. We'd never seen an armory. They told us to get lost or else, so we drove around behind the place and parked the car. The fences were 10 feet high with lots of greenery growing around it. I decided that I would take a shot at shimmying up the fence and have look. Those signs bothered me a little as *My buddy#2* gave me a boost up. There was a neighborhood guy who was walking his dog and was curious to see what we were doing. I jumped down when I saw him, we didn't need the locals ratting us out to the ANG. Everything from what I could see was under tarps. There could have been a helicopter in there but we couldn't be sure.

We headed back across the Verrazano Narrows Bridge and headed back to the city and 12th Ave. It was almost 3:00pm when we arrived at *Evil I. My boss* wasn't thrilled with the report; he was getting more worried, just as we were.

Just about a week went by when on afternoon, I got a phone call from *My boss* at my Universal office. He asked me if it would be possible for him stop by, he needed to talk. I was a little surprised by the call, but he sounded upset and nervous and he didn't want to talk on the phone. We all knew that the phones at *Evil I* were tapped. It was amazing how many times conversations you had with somebody suddenly turned up all over the place a couple of days later. I told *My boss* to come on over and then called *My buddy#2*. and *my buddy #1* to let them know what was going on. *My boss* showed up at Universal in record time. He really didn't look too well. I showed him into my office, closed the door and we sat down to talk. *My boss* started talking immediately about the helicopter situation and how nervous he was about it. He couldn't sleep at night, he said he was worried we could all take the rap for this.

It was obvious to me he wanted some kind of help. There wasn't much I could tell him, except we made a concerted effort to find the helicopter. As to the rest, well if anyone found out? What I said wasn't good enough. I sat for a minute and thought. *My boss* really did look quite upset. There was only one more thing to do, and that was to call the GSA and tell them the truth, the Huey was MIA. I explained to *My boss* that once I made that phone call, there was no going back.

It was going to blow up like nothing any of us had seen. *My boss* sat there and shook his head yes. I repeated, " You want me to make this call? What about *My buddy#2* and *my buddy #1*? Shouldn't they know? *My boss* asked me to call the guys and make it a conference call. The guys agreed. We did make a concerted effort to find the Huey and why should we be the ones to take the rap- when she was discovered later on to be missing. Okay, so I made the call to **PL** at GSA, with *My boss* staring at me as I did just that.

PL answered with the usual cheery greeting. It got pretty quite in the other end of the phone when I starte*d* telling him what was going on. I explained that in our last conversation that I had misled him into believing the Huey was okay. I explained that I had done it to protect the museum and that I was caught off guard. *My boss* just sat and listened. **PL** pulled up the paper work on his end and looked at the loan agreement. I knew that none of us had signed it, so someone had to have signed for the Museum. We just didn't know who it was because we couldn't find the contract. **PL** then told me that he would have to report this to the Inspector General. **PL** also felt that because we had been up front and reported it, he reassured me that we were doing the right thing. GSA would protect us as much as they could.

I hung up the phone and felt really sick. *My boss* was looking a bit more relieved about things, but I just felt ill. *My boss* thanked me and after talking a bit more, it was already 5:00pm and we got ready to leave for the day. On the train ride home, I really started to wonder where all this was going. Things were doing okay at the ship. #122 was getting all the attention she needed. Of course, we had the usual battle of wills with OPS but on the whole, we were doing okay. I didn't know what was going to come next.

Chapter X: Threats and a Change of Command

A few weeks went by after the phone call the GSA. We hadn't heard anything until one day *My boss* got a call from the gate guard at the Museum. A gentleman by the name of **AV** wanted to see him and the gentleman had a badge. It turned out that the GSA NY division had been sent down to *Evil I* to inquire about the MIA Huey. He seemed nice enough *My boss* said when we spoke later. Museum Director was stunned by the appearance of the GSA guy. Museum Director, from what *My boss* said looked like he was pretending he didn't know what was going on. I wasn't there so I do know how that went down. I do know that Museum Director did have to call *Ace* into the office and the sweat was really flying.

From what I gleaned from *My boss*, it wasn't pretty. GSA had caught *Ace* boy pretty much unawares. When GSA showed him the copy of the contract *Ace* had signed, My boss said Museum Director went white. It was obvious that *Ace* boy had overstepped his authority and signed for this helicopter and apparently the deal was made with another GSA guy in NY by the name of **PC**.

PC had okayed the request for the Huey. Why? Because, it was going to a VFW as an exhibit. You name me any VFW that has a Huey 1-M in its lobby. When *Ace* was asked where the VFW was, *Ace* just said that it was for his brother's VFW. *My boss* said that *Ace* was also asked was it the VFW in Staten Island. Ace refused to answer. Next thing that I heard was **AV**, then pretty pissed off, told everyone at *Evil I* no to go anywhere until he got back. He was on his way to Staten Island to find the MIA Huey. I know this to be a fact because *My buddy#2* called me to ask for the directions to the place we went to. *My buddy#2* later told me that *Museum Director*'s office door was closed but you could hear the screaming in the shop that's how loud it was. *My buddy #2* just kept his head down and stayed out of the way. *My boss* later told me the secretary for the *Evil I* board of directors *WB* was also finding his way to *Museum Director*'s office to join in the fun. *WB* was just as guilty as anyone there was . He used *Evil I* to fund his little weekend jaunts to Miami, running up 1000 dollars worth of cell phone calls during the trip. He was a slag who used *Evil I* to fund his "games" and "buddies". He should have been hauled out in cuffs years ago. Maybe had the things that went wrong with *Evil I* wouldn't have if he was out of there. (to close this, he was eventually out of Evil I after paying the Attorney General a lot of money for some sort of trading deal)

My buddy#2 called me later to tell me that Anthony came back and told everyone there that he had found the helicopter, right where he figured it would be, under the tarps. *My buddy#2* later told me that **AV** had to threatened to get GSA agent and the cops down there before they finally let him into the place. He walked to the garages and right where I thought I saw the helicopter, under those tarps is right where he found her. *My buddy#2* said that as he was leaving, he saw Ace in the hallway and he was looking bad, real pale and nervous. He didn't even look at *My buddy#2*, just kept walking. To say the least by five that afternoon things had definitely changed on *Evil I*. *My boss* was laying low. He had been summoned into *Museum Director*'s office to tell what he knew about the helicopter, which he told them he knew nothing.

The same question was being asked, how did this get out. Someone from inside the *Evil I* establishment had to have told someone outside. *Museum Director* obviously knew something, but decided to turn his head because of the fact *Ace* was into the big guys on *Evil I*, people like **WB** who let him get away with anything. We had not heard the last of this, not by a long shot. Change was in the air and I knew it wasn't going to be good.

In the weeks following the GSA raid, the 1-M helicopter was picked up by GSA and taken from the VFW in Staten Island to a vocational school inn Long Island to be used as a project for students. Well, at least she was safe and would get a nice restoration job on top of it. It wasn't the end though. *Ace* and *WB* and even Museum Director, his buddy, were persecuting *My boss*. *My boss* was now being left out of meetings, passed over for raises, ignored and just about told he was on the way out. *My boss* even went thorough this "grilling" where *Ace* or *Museum Director* or *WB* would batter him until he would tell how and where the GSA found out about the 1-m. *My boss* never budged, he just clamed up and gave nothing away. In the meantime, we tried to make it life as usual on the flight deck, but it wasn't easy. Things were always going wrong for us. Usually, it was because someone in OPS wanted it that way. They were convinced it was us who gave away *Ace's* brother's helicopter.

During this siege of the helicopter, there was another project that we were working on. I think we were all so screwed up that none of us was thinking clearly. If I had been I would have NEVER gone as far as I did with this new project, the F-8 Crusader. We were in the middle of trying to get her acquisitioned to come to *Evil I*. Pensacola Naval Museum was giving us a bad time about getting it.

The aircraft was in Warminster, PA and Pensacola was bitching that we were undercutting them with the Naval Curator's office because we were closer than they were. *Museum Director* wasn't exactly helping us along, he seemed preoccupied with something else. We were to find out that something else would be the fact that there was going to be a new CEO (something *Evil I* never had before) coming in. *Museum Director* was no longer the sole director of the operations as he had been for something like 15 years. This new guy was a big secret, but the word was out that he was a Marine General. Great!! Just what we needed, a retired jarhead running the show! He also showed up just as we were trying to get an answer from the board on whether we had permission for acquiring the Crusader.

General **S** came onto *Evil I* in a big deep dark secretive way. No one saw him for about two weeks after he arrived. And he arrived in style, limo and all. We found out that he was a retired marine general, had some sort of financial background and also as the words was going around, had been told he had to 'leave" the marines or be prosecuted for an insurance scam that the marines found out he was running. I don't know where the proof of that is, but it was the local story.

He retired and took the job here. It didn't take much to figure out why he got the job. There had been some 20 interviewees for the position. Admirals and Captains in the Navy were some of them. We also heard that the Navy was dropped out of the race because the board wanted someone who would take orders and not question, which was the modus operandi of the Marine corp. that's what Marines did best, take orders and didn't ask any questions. Well, here we were. **Navy Z** down at the Philadelphia Navy Yard wanted a decision, did we or did we not want her. This was to be our first meeting with the General. *My boss* called me up at work one day and said that we had until 10:00AM the next morning to have a full report as to why we should have this aircraft. WHAT?! I said, that's not enough time. *My boss* told me that was all the time we had and could I do it. Yeah, I could do it. I called down to the Naval history office and begged a favor from one of my buddies down there to get me the aircraft's entire history. He faxed it to me, 50 pages long. With that, I wrote the reasons why we should have this historical aircraft, how we were going to do this, how we would get her here and the fact that it would cost us not one red cent. *My buddy#2* had been working with the USAF ANG in New Jersey that were a heavy lift operation complete with a Chinook heavy lift helicopter. The trick would be to pick her up in Warminster and bring her over N New Jersey and over the Hudson and onto *Evil I*'s deck.

I outlined the whole show for General **S.** I did a lot of it sitting in my Orthopedist's office waiting for my appointment for my back. To be honest, I didn't hear a word he said, and he knew it too. He advised me to get some stress management. He was lucky I wasn't paying too much attention to him because I would have knocked him out right there!

So, I went home and between phone calls from My buddy and My boss all night, I managed to get this thing done at about 3:30 that morning. My buddy picked it up at 7 on his way to the ship, and I went to work, exhausted. My buddy called later and said that *My boss* got it onto his desk at 9:50 AM. My buddy also told me that if we didn't have it there by 10, it was a done deal because General **S.** would not support us and *Museum Director* had nothing to say about it.

I couldn't believe that this was happening to us. From all of the other stuff I heard, this General S a was fruit cake. *My buddy#2* relayed a story that *My boss* had told him about being in a meeting with this guy and hearing him rant on about nothing for 15 minutes then turn around and ask all of them for their support.

Sounded like a regular jarhead to me, as far as I was concerned. So anyway, we now had to sit back and wait for his stupidness to pass judgment on our project. We didn't have to wait too long, because *My boss* called me to tell me that *S* had approved the project and he was really impressed with the report that we had done. *My boss* then told me the General wanted to meet me. He had heard all about me from *Museum Director* and the guys and he wanted to know who I was. Just what I wanted, I couldn't freaking wait.

I showed up at 1:30 pm a couple of day s later for my 2:30 appointment with General *S*. My back was killing me, I felt like garbage and I was in no mood for anything. *My boss* met me on deck and told me how General *S* was so impressed with the report and how it was done and wanted to know how I did that so fast etc. Yeah right. I was standing by #122, gave her belly a good rub and then went downstairs to his office. He was cordial enough, this oversized, white haired, red nosed, obnoxious guy. I sat down and before I could get to say one word, he started reaming me as to why I didn't tell him about the missing helicopter.

I looked at him like he was nuts. First off, I said, I didn't know who the hell you were. I'd never met you and you weren't even on board yet. Why should I tell you anything? That shut him up for a second. The second barrage, you should have told me what was going on, I depend on my people to keep me informed. Hey bud, I'm not your people, I thought. The guys were right. This guy was a fruitcake!!

I think he was looking for some sort of reaction, like I was going to get upset and cry or something. This idiot didn't have a clue what he was dealing with. I could out stare Mt Rushmore, he wasn't about to upset me. The next barrage consisted of how he wanted team players and he had heard about all that I had done here and he wanted me to stay on and help him. And IF I DIDN"T LIKE IT I COULD GET OUT!! He screamed. I swear to God, this is just what he did. I just sat there. Then he comes out with, I like your style. "You're Air Force trained aren't you, I can tell".

Yeah, I was Air Force trained but you weren't going to hear about it. At that point, I'd had about enough. I just stood up and told him that this was nice, but I didn't want to take anymore of his time.

I felt like I was in the middle of Marx brother's movie. He stood up and grasped my hand and told me he was looking forward to working with me and he was anxious to see how we were going to handle the incoming F-8 Crusader project. I thanked him and left the office. I met the guys on deck and told them that he was certifiable, on which we all agreed. The Crusader was just about 3 days away and I'd worry about the rest later.

The big day had arrived and with it so was our F-8 Crusader. *My buddy#2* called me by cell phone. He was on his way down to Warminster to meet up with the guys from New Jersey and their big lift Chinook helicopter. *My buddy#2* was going to fly back with them and the F-8. I was going to be waiting with **ND** on deck for the delivery and reception. Of course, the day was beautiful and General **S** was there with some VIP's driving me nuts. I was trying to get the deck cleared and ready, **ND** was on the radio to *My buddy#2* and OPS was being the usual pain in the ass, not to mention totally useless. The sight coming over the river was awesome. Out of nowhere, all of a sudden here comes this gorgeous F-8 attached to a cable and hanging underneath the Chinook. She was hauled over the Hudson River, stopping the traffic on 12th Avenue below. People were climbing out of their cars to see this amazing site. And it was truly amazing. The guys from the heavy lift unit National Guard/Warminster PA, did an awesome job of setting her down gently on the deck, then backing off, releasing the cable and turning to land further down the at the aft end of the flight deck. It was a breathtaking site and experience! Our F-8 Crusader had arrived in style and she was beautiful!!

Chapter XI- The end of the line and the onset of Ravens:

Shortly after the F-8's arrival, we *started* to see some very disturbing things going on at the ship. First off, people were being laid off left and right, money was being cut from the aircraft maintenance fund. The word was the **S** was going make the museum pay for itself and he didn't care how he did it. I want it noted here that **S** was responsible for decimating the entire curatorial and restoration staff of the museum, so he could save money. I also want it noted that it was clear that he was also cleaning house. I got a call from *My buddy#2* one spring day to tell me he no longer worked at *Evil I*, he had been laid off. **ND** had also been laid off, and others were to follow.

There were some very upsetting and tense days at the ship. I knew that there was nothing more that I could do for #122. The atmosphere there was sick. The next to go was *Museum Director*. This is the guy who gave his house up so that *Evil I* could function. He second mortgaged his house, lost it along with his first wife for the sake of the museum and they cut him off like he was a piece of bad cheese. *My boss* soon couldn't take anymore and he left.

I also left. I didn't have a choice I could no longer support this sick regime on the ship. I cut up my ID card and sent that along with a letter of resignation to General **S,** the letter wasn't pretty either. It told him exactly what I thought of him and the rest of the *Evil I* slime. I went back to the ship one more time after that, just to say farewell to #122 and the girls. It was so sad that I don't care to remember it here. It's too much to bear.

All that I had managed to accomplish, was destroyed by the creeps that felt nothing more for *Evil I* than what they could suck off of her for their own pockets. I got very sick after all of this. I watched *My buddy#2* go into a depression and not care enough to look for work. I wrote letters to the Navy curator's office telling them exactly what was going on there. I even went with another *Evil I* Casualty to the Attorney General's offices to show them the inconsistencies that were going on at *Evil I*. All of this was to no avail. I felt it so hard to let go. I even went to the newspapers and got a journalist to write an expose on what was going on there. **S** went to his buddies on the NY Times and got them to refute it and say we were just disgruntled employees.

It went back and forth for some time. I did come out of there with one thing. I made a promise to #122 that JV and I would write her story. At least we did do that much and it was a success, albeit MBI tried their best to drive us totally mad.

I did go back to the ship and attempted to re-volunteer at some ridiculous point in time, only to be found out by one of *Evil I*'s VP's, **JR.**

This is good, the guy didn't even have a high school diploma or so I was told, he certainly didn't have the credentials to be an exhibits curator and they give him a position of VP of exhibits. I don't know about you, but I've never heard of any museum having a VP of exhibits, especially one that doesn't even have a curator!

Evil I herself was turned into a floating nightclub with the pretense that they are some sort of museum. I'm not sure what kind. *My buddy#2* and I have both had our lives threatened if we ever showed up on the ship again, not that that ever stopped us from visiting the aircraft when we could. The only good thing that did make us feel a bit better is that the Department of Defense Special Investigations department showed up at my buddy#2's front door and they also came to my office to talk to us about what went on there. I did hear that one member of the *Evil I* thugs did get his comeuppance.

As to #122, I still missed her deeply as I missed all my aircraft. I still hope that one day, all of the problems at *Evil I* will go away and once more she will become the Museum that she should be.

Chapter XII: #122

I did the best that I could. I gave everything I had and more. It shouldn't have ended like this. I still believe that things happen for a reason and one day, it will all be resolved. Let me close this story with an incident that happened right after we had left the ship for good as volunteers and employees. It gives me a reason to believe all of this and #122 will one day be resolved. My buddy#2 and I had snuck onto the ship one day to visit. It was early in the morning; the museum had just opened. We had gone up to see #122, and we're just standing up on the deck, almost ready to leave. We never stayed too long, for fear that we'd have to have some problems with the idiots there, I was turning on my cane to leave when above me I heard some, what I thought to be crows. My buddy was standing just in front of #122 and I had a camera with me, standing at her side. A huge black raven landed on #122's nose and proceeded to scream at us both, loudly. I snapped a photo of it. Neither one of us could believe our eyes, so the photos came in handy later. We could see it again. I like to believe that was #122 saying keep up the fight. Keep screaming, you will win.! That's what I want to believe and that is what I will believe.

Epilogue: Today

Well, here we are 20+ years after the fact. #122 is still on *Evil I. Evil I* has actually undergone a change for the better. Perhaps it's more a commercial venue than a museum, but better. She has now acquired a space shuttle as the main meal ticket.

#122 is not so much in the spotlight anymore. Much of the flight deck has been taken over by the balloon tent that covers the shuttle exhibit. Some things never do change. During Hurricane Sandy, the wisdom of putting a priceless shuttle on the flight deck of an aircraft carrier 17 stories above the Hudson river was totally stupid. The balloon tent came down in the horrific winds of that storm and injured the shuttle, which I still understand they have not been able to repair properly given there are no shuttle parts available. It's just another example of the lack of wisdom when planning for exotic aircraft in this museum.

However, there is some care being taken to look after #122. I know many people in the aircraft conservation industry and I get the word on what is going on where. They are still trying to raise funds for a painting of #122 which I find hard to quantify, given that should be part of the financial spreadsheet of any aircraft museum.

I am now the "go to person" for the Lockheed A-12 for the "Roadrunners Internationale Association" of which I am a honorary life member. Roadrunners Internationale comprises all those wonderful, "silent warriors" that worked on the U-2, A-12 and SR-71 programs at Area 51 during the Cold War. Together, we do keep an eye on all the blackbirds that are out in the museums.

My partner JV and I started Phoenix Aviation Research and wrote the first, complete book on the A-12 and followed it up with a new book coming out the summer of 2017 called :*"Black Lightning- the Legacy of the Lockheed Blackbirds".* It was the book we always wanted to do on the OXCART program and the blackbirds.

As things have it, time has passed, #122 is still viable, which is totally amazing considering what that airframe goes through sitting on that deck in all kinds of New York weather. The real place for her is back at the National Museum of the USAF where she would complete the only showcase of all the Lockheed Blackbirds. Who knows, we just might be able to pull that off some day, you never know.

I never thought I would put this story down on paper, but I have and sincerely I love and thank all the people that I met and worked with and on the Evil I. Some are now life long friends. *My buddy #1* is an executive with a large aircraft manufacturer and my dearest *buddy #2* is a VP at a College of Aeronautics. I am so proud of him. He travels worldwide for the school and never forgets just where he came from. He is amazing.

I still love *Evil I* deeply and wish her well and long life. She should always be honored as the aircraft carrier that fought WWII, Vietnam and picked up astronauts on the way home from space. She has a magnificent history. Both her and her crews should be honored for their service.

If anything comes out of this story, it just proves no matter what, if you love what you do, if you really believe in what you do, you can accomplish it. I did, I still can't believe I have been so lucky to have been able to live out my passion for airplanes and to have served to save a vital piece of aviation and Military Cold War history.

I have been blessed in having a partner like JV who believed as I did, that all things are possible if you care and love it enough.

Between both of us we wrote the only book dedicated solely to the Lockheed A-12 and the OXCART program. We took on the CIA by use of FOIAs, research and some real fast "okie-dokeing" of our own and managed to right the confusion concerning an A-12 pilot that was not given the credit deserved for a very special mission. We were pleased to be able to give his family his whole story and confirm he did make the flight that found the USS Pueblo in 1968 after she was taken prisoner by the North Koreans.

I ended up with a Master's degree in Aviation Science, a number of Aviation history books that I wrote on the shelf and continue to write history and paint aircraft portraits for the USAF Art Program where I was honored to be included in a most beautiful book, "Magnificent Showcase: The History of the U.S. Air Force through Art" by Dr. Timothy Keck.

I truly hope that this story of what we did to save an aircraft that is so unique to the Cold War, will help people to remember those who served and won the Cold War, the pilots that were lost flying the A-12, and all those who worked and flew the A-12 in those long hours at a place that didn't exist (Area 51). That is really what saving #122 was all about. Future generations can look at the A-12 and know about those people who gave their all in silence to win the most dangerous war......The Cold War.

Remember them when you look at A-12 #122 or any of the Blackbirds. That is why saving this aircraft meant so much. She wasn't just an airframe, she was and is living history.

#122 is a monument to the genius that brought her to life (Kelly Johnson and the crews of Lockheed) and those that served with and on her. The Blackbird spirit still lives and she will always be the fastest reconnaissance aircraft in the world, at least in our hearts and those of her many Blackbird fans and those who worked the program.

                              Blessings,
                        Jeannette Remak
                          (aka, Det 1)

www.ingramcontent.com/pod-product-compliance
Lightning Source LLC
Chambersburg PA
CBHW071439180526
45170CB00001B/390